Editor
Karen Tam Froloff

Managing Editor
Karen J. Goldfluss, M.S. Ed.

Editor-in-Chief
Sharon Coan, M.S. Ed.

Illustrator
Bruce Hedges
Blanca Apodaca

Cover Artist
Wendy Roy

Art Coordinator
Denice Adorno

Imaging
Alfred Lau
James Edward Grace

Product Manager
Phil Garcia

Publishers
Rachelle Cracchiolo, M.S. Ed.
Mary Dupuy Smith, M.S. Ed.

PRIMARY

Author

Mary Eggers

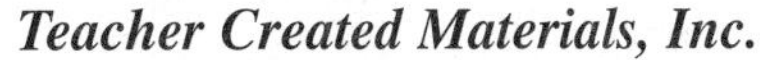

Teacher Created Materials, Inc.
6421 Industry Way
Westminster, CA 92683
www.teachercreated.com

ISBN-0-7439-3301-X

Table of Contents

Introduction 3

How to Use the Book 4

Expert Math Detective Badges 6

Problem—Solving Stories 7

Book 1: The Dino Disaster 7

Book 2: The Treasure of Nog 19

Book 3: The Alien Invaders 31

Book 4: The Haunted House 43

Book 5: The Insect Investigation 55

Book 6: The Scuba Scare 67

Book 7: The Future Flight 79

Detective's Kit (Manipulatives) 91

Introduction

What are Problem-Solving Math Mysteries?

Children love a good adventure story. It is even more exciting if they are "living" and experiencing the story! This book is a collection of seven math adventure books. The student becomes a "math detective" and solves a series of story problems or word problems in order to finish each adventure.

The story problems in these books are created from the math skills taught in the primary grades (addition, subtraction, simple multiplication, division, money, time, fractions, comparing numbers, measurement, symmetry, number patterns, perimeter, and shapes).

Several pages of manipulatives (called a "Detective's Kit") have been included at the end of this book to assist the student in solving the problems and can also be used by the adult when teaching some of the math skills to the student.

Each book is divided into several parts. Every part has an accompanying math problem that must be solved before the student can move on to the next part of the adventure. Five answers with five page numbers are given with every problem. If the student chooses the correct answer, a page number is given that continues the adventure. If the incorrect answer is chosen, a page number is given that describes a less desirable turn of events such as being chased by angry dinosaurs. Then the student must return to the last problem and try again.

Work must always be shown to ensure that the child can work and understand the math problem. (**Note:** If students need more work space they can use the blank page that precedes the page to which they must return.) When the adventure is successfully completed, the student is given an Official Expert Math Detective Badge and promoted to the next detective level. *Problem-Solving Math Mysteries* connects math and reading while reinforcing critical thinking and problem-solving skills. In addition, this book shows the student that math is not only useful but can be lots of fun!

Who Should Use Problem-Solving Math Mysteries?

The stories in each book are written for the primary grades and can be used in a whole-class setting, as an enrichment or supplemental activity for individual students, or at home.

It is recommended that if the books are being used in whole class instruction at the first grade level, they should be introduced later in the year after the students have been taught all of the math skills addressed in the books. It is also recommended that the teacher have the students work in teacher-selected groups of four that each contain a strong reader and a strong math student (or a child with good problem-solving skills) for optimum success. The students should each have their own story book (detective's notebook) and should each be responsible for showing their work. Remind them that they will have the assistance of other group members to figure out what is being asked and how to solve the problems. It is also beneficial to assign jobs to each group member such as the following:

1. **Reader** (This student reads the story parts to the group.)
2. **Recorder** (This student can record notes/ideas for group on scratch paper until the final answer is found. Then all members record the work in their individual books.)
3. **Task Master** (This student makes sure that everyone is participating and on task.)
4. **Questioner** (This student is the only one allowed to raise his or her hand for adult help and then only if no one in the group can answer the question.)

When using the books in whole class instruction with second and third grade students, the stories can be worked any time during the year in small groups, pairs, or by individual students (especially in third grade), depending on the ability level of the class.

How to Use the Book

Assembling the Stories

Seven adventure mysteries are included in this unit. Each book is 24 half-pages long (12 full pages). Once a book is photocopied, it can easily be cut along the dashed lines, collated, and stapled. Each book has illustrated pages, and each of the pages is numbered. The books are also numbered (1–7) so the adult knows which story should be used first since the later stories have more story/word problems than the earlier ones.

It is recommended that three staples be used on the left-hand margin because the pages will be turned often since the problems are scattered throughout the book, not in chronological order. (That would be too easy!)

The manipulative pages (Detective's Kit) must also be photocopied. Some of the pages include objects that need to be cut out (clock, coins, ruler), while others are reference pages (pictures of shapes, 100's chart, etc.). It is most convenient for each student to have his or her own "kit" so that it may be taken home and shared along with the books once all of the cases have been solved, but it is possible for a "kit" to be used in pairs or a small group, if necessary.

The "kit" pages can be stored in folders that the students may decorate since the folders will be needed for all of the stories. Before storing manipulatives in each folder, place them in an envelope or sealed plastic bag. Collect the folders between uses so none of the pages get lost.

Reading the Stories and Solving the Math Mysteries

Before students begin the first book, model how to use the books by following these steps:

1. *Set the scene for the stories.*

   Explain that the student is a junior math detective on his or her first case. The student detective will be embarking on a special adventure. They will have a detective's notebook (the assembled story book) that tells their story and must be used to show all of their detective's work. (This is best done on the blank back of the page opposite the problem.) They will also be given a secret detective's kit of hints and tools that will help them to succeed on their mission. If they can solve each problem in the story successfully, they move on to the next part of the adventure and ultimately solve the case, get promoted, and get assigned a new mission. If they cannot solve a problem, something unfortunate will happen to them and they will have to try the problem again!

How to Use the Book *(cont.)*

Reading the Stories and Solving the Math Mysteries *(cont.)*

2. *Introduce the Detective's Kit (manipulative pages 91–96).*

   Reproduce pages and demonstrate how to use the items in the section (number lines or 100's grid for number pattern problems, multiplication table for multiplication and division problems, etc.).

3. *Model how to show the work in the detective's notebook (the storybook).*

   An answer alone is not sufficient. To demonstrate understanding of the problem and its solution, the student must draw a picture, write an equation, or write a simple sentence such as "I got the answer by using the clock model in my detectives kit." Remind students that they must show proof that they know how to solve the problem just like real detectives do when they file a report.

4. *Work the first two or three problems with the students.*

   If students have not been exposed to many story or word problems before, this becomes a teachable moment. For a whole class lesson, make an overhead transparency of the problems.

Here are several clues to share with the class. You may wish to post them in the classroom as reminders for students.

Detective Clues for Solving Math Story Problems

- ❑ Every problem should be carefully read two times before trying to solve it.
- ❑ Look for words that give you hints on what is to be done such as "all together" or "in all" which tends to mean you add, and "got away" or "loses" which tends to imply subtraction, etc.
- ❑ Use the detective's kit whenever possible.
- ❑ Read all of the answers before choosing one. At least one or two should obviously be wrong. There are also a few problems with more than one correct answer or an "all of the above" option.
- ❑ If you are really stuck, try working backwards. Choose an answer that looks reasonable and try to prove it correct. (Remind the student, however, that a picture, an equation, or a sentence showing how the correct answer was arrived at must accompany all answers.)
- ❑ Use a bookmark or self-sticking note to mark the page you are on so that you do not lose your place if you turn to a page that is not the correct answer and you have to go back and work the problem again. Remove the bookmark only when your work is shown and you have definitely found the correct answer and the next part of the adventure. Then place the bookmark on this new page and continue as above.
- ❑ **When you have found all the correct answers, highlight them on each correct page. Then, read the story from beginning to end using only the correct pages.**

Good luck on your adventure!

Expert Math Detectives Badges

Reproduce these badges on heavy paper. Make enough copies for each student in the class.

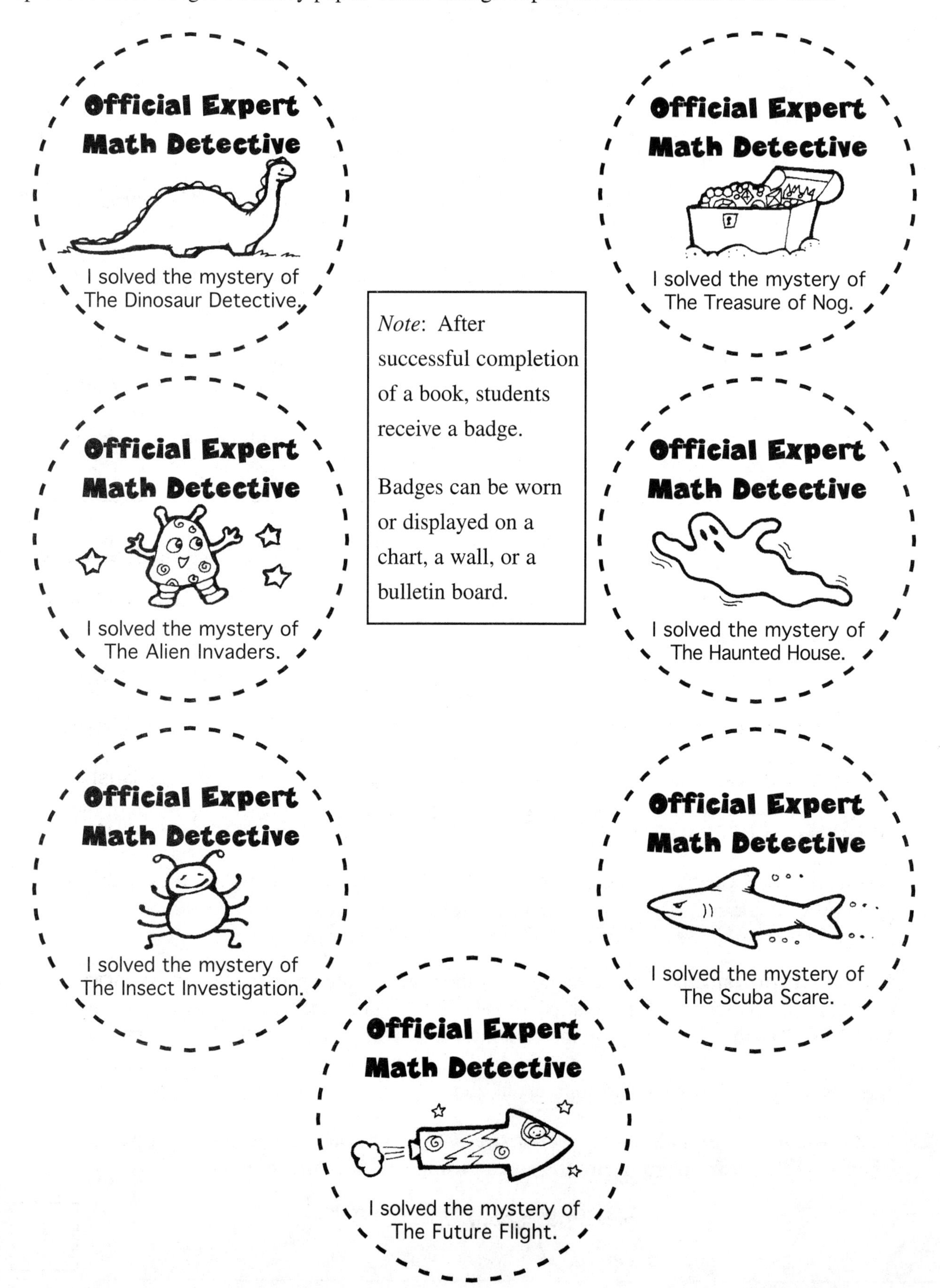

Note: After successful completion of a book, students receive a badge.

Badges can be worn or displayed on a chart, a wall, or a bulletin board.

The Dino Disaster

Detective's Name:

____

Date:

____

Hello, Junior Math Detective. Good luck on your adventure, and don't forget to show ALL of your work in this book. Have fun, and...

DON'T BECOME DINO DINNER!

You and your friend are helping to clean out your uncle's shed when you find a strange chair. It is made of metal and has a clock and colored buttons on it. Your friend sits down on it and starts pushing buttons. There is a bright light. The chair vanishes and returns EMPTY! You know you have to try to find your friend. You sit on the chair. The last button that was pushed was a three-dimensional (3-D) round shape. You must push it to find your friend.

If you choose... a pyramid, go to page 2.
a cube, go to page 4.
a cone, go to page 6.
a circle, go to page 3.
a sphere, go to page 5.

"Get away from my shed!" your uncle yells. This is not the right answer. Go back to your last page and try again.

2

Oh no! The chair falls apart. Now you'll never find your friend! This is not the right answer. Go back to your last page and try again.

3

Yipes! The chair turns into a lion, and you are sitting in its mouth! This is the wrong answer. Go back to your last page and try again.

4

You feel the chair shake. You close your eyes. When you open them, you see that you are in a jungle. You don't see your friend, so you jump off the chair and look down. Someone wrote in the dirt with a stick. It's a clue from your friend and says, "Add 10 and 8 and 2 and 3, then take away 5 and you'll find me!" You hope that you can do this tricky problem.

If your answer is 18, go to page 7.
17, go to page 3.
19, go to page 10.
28, go to page 6.
27, go to page 8.

5

Help! The chair got impatient and left without you. You are trapped! This is not the right answer. Go back to your last page and try again.

6

"Eighteen!" you say. Then you notice an 18 with an arrow carved in a tree. You are just about to follow the arrow when you hear a loud cry and feel yourself being lifted into the air. You gasp and look up. A huge Pteranodon has you in its claws. It drops you in a nest on a mountain and flies away. It is a big nest! One side is 3 feet long, two sides are 4 feet long each, and the last side is 2 feet long. What is the perimeter of the nest (or, how many feet around is the nest)?

If you choose...

9 feet, go to page 11.
13 feet, go to page 9.
10 feet, go to page 13.
12 feet, go to page 8.
24 feet, go to page 10.

7

EEK! A huge rainstorm pours down on you, and you have nowhere to hide! This is not the right answer. Go back to your last page and try again.

8

That's one big nest! You get out and begin climbing down the mountain when you hear a noise. You turn around to find that you are looking into the open mouth of a Tyrannosaurus rex! You grab at the bushes growing out of the cliff. If the T–rex can hold 50 pounds of food in its mouth at one time, and each bush weighs 5 pounds, how many bushes will you need to throw into the T–rex's mouth to fill it up?

If you choose...

45 bushes, go to page 8.
5 bushes, go to page 11.
20 bushes, go to page 13.
55 bushes, go to page 10.
10 bushes, go to page 14.

9

Yikes! A meteor is zooming towards the earth. You'll never have time to find your friend before it hits! This is not the right answer. Go back to your last page and try again.

10

Help! You start to run away, but trip over some Maiasaura eggs and the parents DON'T like it! Here they come! Keep running! This is not the right answer. Go back to your last page and try again.

11

"We'll never make it to the chair by 12:00," your friend gasps. You'll have to trick the T-rex. You duck behind a tree as it roars past. Whew! Then you see a baby Velociraptor coming towards you. It's pretty small, so you and your friend decide to throw something at it to scare it away. You check your pockets. All you have are a few coins. If you throw 1 quarter and 2 pennies, and your friend throws 1 dime and 2 nickels, how much money did the both of you throw in all?

If you threw...

44¢, go to page 16.
30¢, go to page 15.
47¢, go to page 18.
52¢, go to page 19.
48¢, go to page 13.

12

Run! An angry Triceratops is charging towards you! This is not the right answer. Go back to your last page and try again.

13

You slide the rest of the way down the mountain and bump right into your friend! "Run!" you scream as you grab your friend's hand. "The chair is this way!" your friend yells back at you as you run. Behind you, the angry T-rex is getting closer! It is 11:40 now. If the chair's clock is set to return the chair to your uncle's shed in 20 minutes, what time is the chair going to take off?

If the time is...

11:60, go to page 13.
12:00, go to page 12.
11:55, go to page 16.
11:20, go to page 15.
12:05, go to page 17.

14

Huh? You wake up to find that your little sister's plush dinosaur doll is on top of your face. Too bad, it was just a dream, and now you'll never know how it ends. This is not the right answer. Go back to your last page and try again.

15

Yikes! A volcano starts to erupt, and hot lava is rushing towards you! This is not the right answer. Go back to your last page and try again.

Watch out! You've run all the way to the sea. You jump in and start to swim. You feel yourself being lifted out of the water. You are on a 40-foot Elasmosaurus' back! This is not the right answer. Go back to your last page and try again.

17

The baby Velociraptor cries and runs away. Uh oh! Someone heard it. Two adult Velociraptors race from behind some trees. You and your friend start running. You get to the chair and hop on just as the T–rex runs up! Your friend grabs a big bone from the ground and begins swinging it at the dinosaurs as you try to start the chair. One button says "HOME" with these numbers written by it: 1, 3, 5, 7, ___, 11, ___, 15, 17. You must figure out the pattern and push the buttons with the missing numbers before the chair can leave.

If you choose...

9 and 13, go to page 21.
8 and 12, go to page 22.
12 and 6, go to page 23.
10 and 14, go to page 19.
2 and 4, go to page 20.

18

Oh no! The Pteranodon is back and swooping towards you with open claws. This is not the right answer. Go back to your last page and try again.

19

Yikes! The chair takes off with your friend but leaves you behind! This is not the right answer. Go back to your last page and try again.

20

The chair begins to shake just as the T-rex is about to bite it AND YOU! There is a bright light. Suddenly you are back in your uncle's shed. You think that it was all a dream until you see that your friend is still holding the big bone that was used to fight the dinosaurs. You decide to take it to a museum. Hooray! It belongs to a new type of dinosaur never found before.

The scientists decide to name the new dinosaur after you and call it a (an) ________________ osaurus. You become famous all over
(your name)
the world. Great job, Math Detective!

21

EEK! The T–rex grabs you and opens its mouth. "Now I have you!" it screams. "And I am going to make you count every leaf in the jungle!" I hope you like counting. You'll probably be doing it for the rest of your life! This is not the right answer. Go back to your last page and try again.

22

Gulp! The chair takes off, but you land on another planet! This is not the right answer. Go back to your last page and try again.

23

The Treasure of Nog

Detective's Name:

____

Date:

____

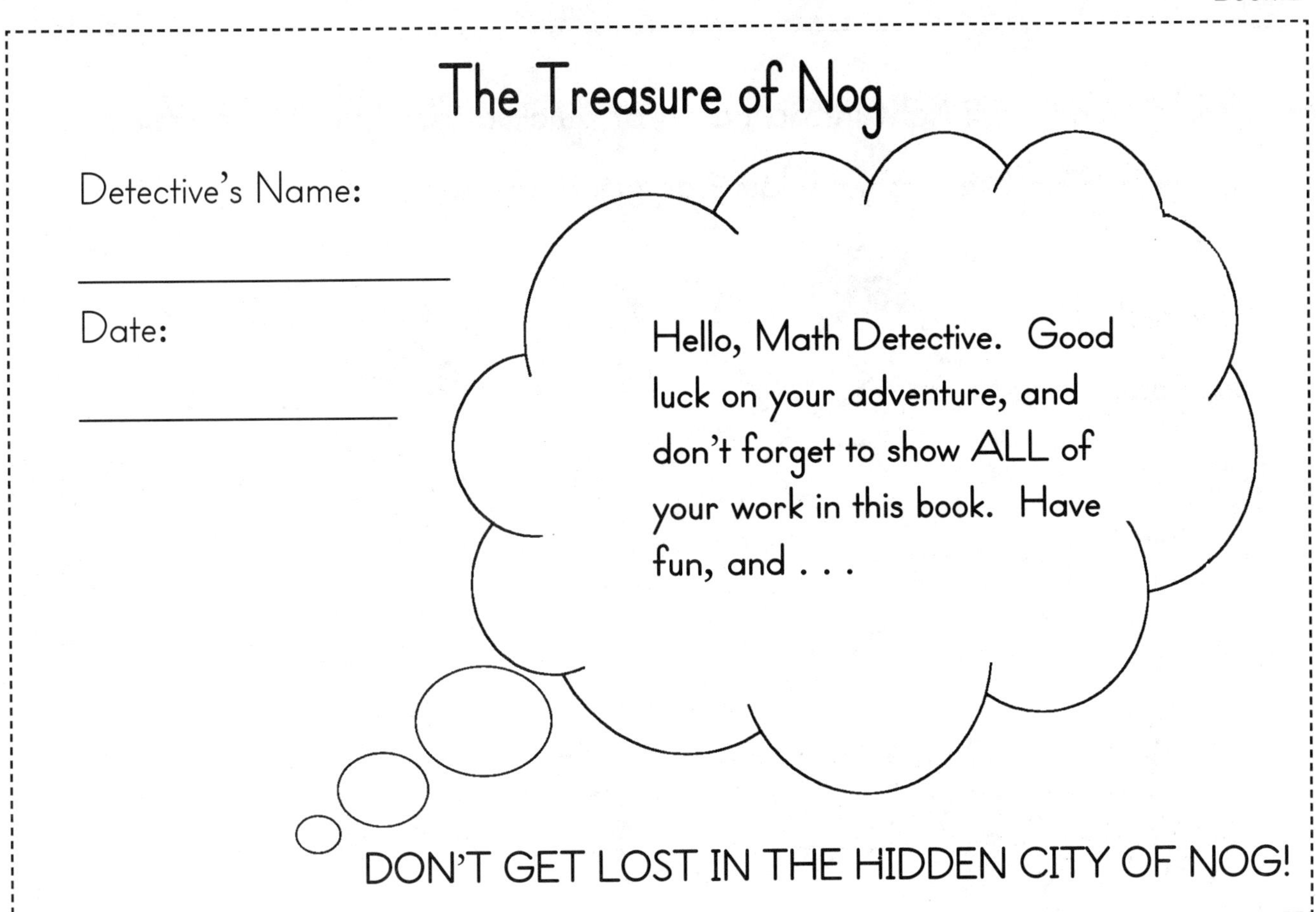

DON'T GET LOST IN THE HIDDEN CITY OF NOG!

You and your friend are walking in a big, dark jungle. You have heard that there is a treasure hidden in the sun temple in the long, lost city of Nog. Your friend is afraid that you are lost. You think that this may be true when suddenly you see 3 old paths. Each has a sign that shows a fractional part on it. To find the right path you must figure out which sign, or signs, show 3/4 as a fraction. (Look at the shaded part of the pictures.)

If you choose... sign #1, go to page 4.
sign #2, go to page 3.
sign #3, go to page 13.
sign #1 and #2, go to page 14.
all of the signs, go to page 6.

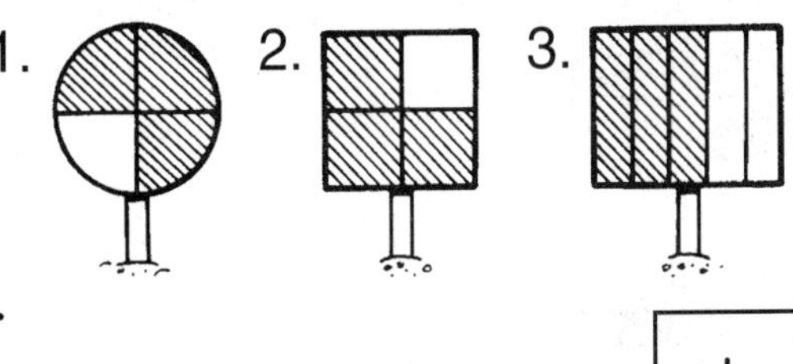

1

Yikes! You walk right into a patch of quicksand! This is the wrong answer. Go back to your last page and try again.

"Almost!" yells a parrot flying overhead. This is not the right answer. Go back to your last page and check all the pictures carefully.

3

"You are half right," a voice whispers from a nearby bush. Go back to your last page and check all of the pictures carefully.

4

You run out of the tunnel into the hidden city of Nog. You see a building with a big sun painted on it. That must be the sun temple where the treasure is hidden! You walk slowly into the temple. You know that there might be traps here. You see numbers scratched on the floor. They say "1, 4, 7, 10, ____, ____, ____." The last three numbers are missing. You see spears sticking out from the wall. You have to figure out the last three numbers in the pattern so you can cross the room without setting off the trap!

If you pick numbers... 11, 12, and 13, go to page 9.

12, 14, and 16, go to page 18.

14, 17, and 20, go to page 13.

9, 8, and 7, go to page 7.

13, 16, and 19, go to page 17.

5

Help! A tiger jumps from behind a tree, and it looks hungry! This is not the right answer. Go back to your last page and try again.

EEK! You turn around and see a huge stone rolling towards you! This is not the right answer. Go back to your last page and try again.

The lid of the box springs open just as your friend slides into the room and runs over. You both look into the box. A big, golden egg is inside. The priceless Egg of Nog! You grab it just as the ground starts to shake. The shaking has awakened all kinds of animals. Huge 3-pound ants and 3-pound spiders are creeping towards you, and a snake with a mouth that can hold 12 pounds of food at a time blocks your path! How many ants and spiders will it take to fill up the snake's mouth (unless it chooses to eat you)?

If you choose... 9, go to page 16. 3, go to page 13.
15, go to page 15. 4, go to page 11.
5, go to page 18.

8

Yikes! An army of huge ants comes running towards you. You are trapped! This is not the right answer. Go back to your last page and try again.

9

Oh no! The ground is shaking again. The tunnel is starting to fall down around you! This is not the right answer. Go back to your last page and try again.

10

You are lucky! The snake likes the taste of spiders and ants more than math detectives. As the snake slides past you, you and your friend jump through a hole in the wall. You are outside of the temple, and what a mess! Buildings have fallen down, and a huge crack has opened up across the ground within a few seconds! Look at the picture below of the crack's size when it began and use the ruler from your math detective's tool kit to measure how small it was in millimeters (mm).

If it measured...

63 mm, go to page 22.
30 mm, go to page 20.
48 mm, go to page 13.
3 mm, go to page 23.
7 mm, go to page 21.

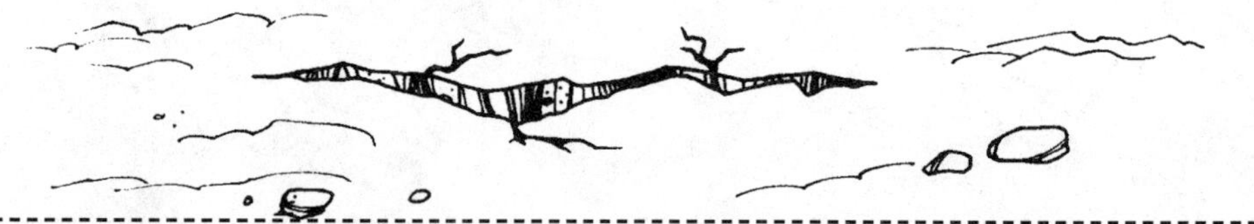

11

EEK! A giant tarantula jumps out in front of you and it looks mad! This is not the right answer. Go back to your last page and try again.

12

Run! Huge flying bugs are zinging through the air at you! This is not the right answer. Go back to your last page and try again.

13

You see that two of the paths' signs show 3/4 fractions, so you and your friend decide to split up and each take a path. You choose the middle path. You walk a long way until you see an old stone wall. There is some writing on it. It says, "When you add 5 plus 6 and 7 and then take away 2, yell out the answer and Nog will show through." What could this mean?

If your answer is... 18, go to page 2.
20, go to page 7.
21, go to page 9.
12, go to page 6.
16, go to page 19.

14

Oh no! The walls are moving. The room is getting smaller and smaller! This is not the right answer. Go back to your last page and try again.

15

EEK! A hole opens in the wall and water starts rushing in. You are trapped! This is not the right answer. Go back to your last page and try again.

16

Whew! You made it. You lean on a wall to rest, but the ground starts to shake again, and the wall gives way. You slide down a hill into a lower room. You land with a clank on a pile of rare coins! The gold coins have "\$25" written on them, the silver coins have "\$10" on them, the copper coins have "\$5" on them, and the tin coins have a "\$1" written on them. You see a box with a hole in it and some writing that says, "Put in \$32 using 7 coins." What coins should you use?

If you choose... \$25 + \$1 +\$1 + \$1 + \$1 + \$1 + \$1 = \$32, go to page 7.
\$10 + \$10 + \$5 + \$5 +\$1 + \$1 + \$1 = \$32, go to page 9.
\$10 + \$10 + \$10 + \$1 + \$1 + \$1 + \$1 = \$32, go to page 18.
\$10 + \$5 + \$5 + \$5 + \$5 + \$1 + \$1 = \$32, go to page 8.
\$5 + \$5 + \$5 + \$5 + \$5 + \$5 + \$1 = \$32, go to page 16.

17

Oh no! The ceiling is moving. It is getting lower and lower. This is not the right answer. Go back to your last page and try again.

18

"Sixteen!" you yell as loudly as you can. Suddenly the ground begins to shake. The dirt caves in under you. You fall into a dark, deep tunnel. You walk along until you come to a pile of dirt. Part of the tunnel must have caved in. You peek over the dirt pile. You see old buildings. It is the long, lost city of Nog! You see an old pail lying nearby. It looks like it will hold 2 gallons of dirt. There are 14 gallons in the way. How many times will you need to fill the bucket to move all of the dirt out of your way?

If your answer is... 7 times, go to page 5.

16 times, go to page 9.

6 times, go to page 12.

28 times, go to page 7.

14 times, go to page 10.

19

EEK! Buildings are falling all around you and you have nowhere to run! This is not the right answer. Go back to your last page and try again.

20

Yikes! A big gorilla runs out from behind a building and begins throwing coconuts and bananas at you. This is not the right answer. Go back to your last page and try again.

21

You and your friend run out of the city of Nog just as it sinks into the huge crack in the ground. You run all the way back home. The golden egg is put in a museum, and you and your friend become famous and are asked to visit schools all over the country to tell about your adventure. Great job, Math Detective!

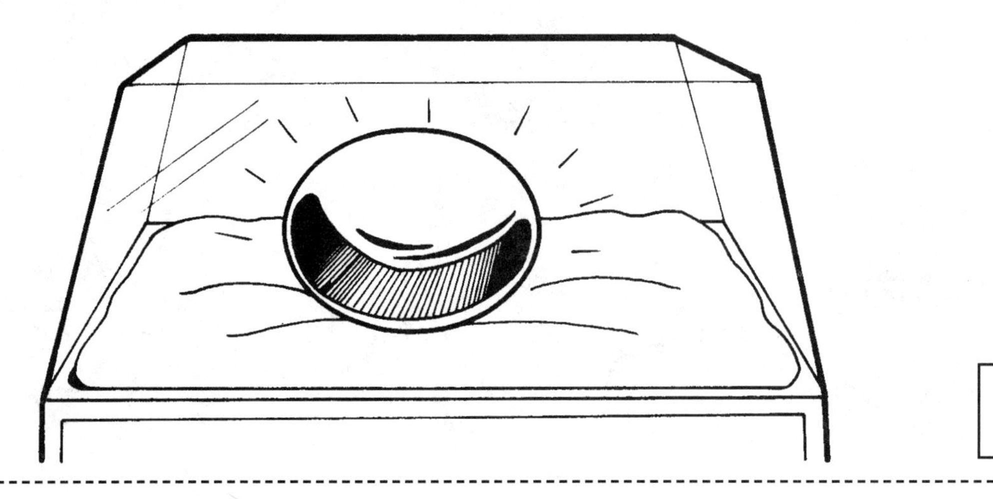

22

Help! A huge stampede of elephants is racing towards you! This is not the right answer. Go back to your last page and try again.

23

The Alien Invaders

Detective's Name:

____

Date:

____

Hello, Senior Math Detective. Good luck on your adventure and don't forget to show ALL of your work in this book. Have fun, and...

DON'T LET THE ALIENS GET YOU!

You and your friend are camping in the woods. It is the middle of the night, and you are trying to sleep when you hear a "THUMP!" You peek out of your tent and gasp! To your right there are 8 eyes looking at you. To your left you see 6 eyes. They must belong to bears that are after your food. How many hungry bears are there in all?

If your answer is. . . 14, go to page 2.
10, go to page 3.
6, go to page 10.
8, go to page 6.
7, go to page 5.

1

Yipes! A bear runs away with all of your food. This is not the right answer. Go back to your last page and try again.

2

OUCH! A bee stings you on your thumb! This is not the right answer. Go back to your last page and try again.

3

Oh, NO! You used up 15 seconds. Only 5 seconds left! You are about to give up when you see a printed time on the door. It says 10:35. There are some cards on the ground with clock faces on them. You grab the one that matches the door and stick it onto the door. It better be the right one. Your time is almost up!

If you choose . . .

 go to page 12.

 go to page 23.

 go to page 18.

 go to page 19.

 go to page 15.

4

You pick up some pots and pans and bang on them. The eyes are gone! You and your friend look around to see if anything is missing. "Hey! Look over here," your friend yells.

You look down at the biggest footprints you have ever seen next to some open bags of food. Which picture shows that the animals ate 1/3 of your food? (Look at the shaded part of the pictures.)

If you choose . . .

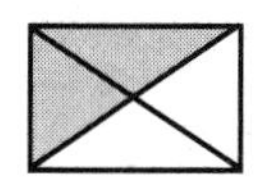 go to page 10.

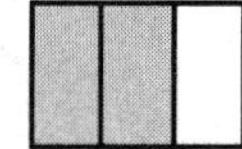 go to page 7.

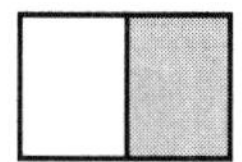 go to page 21.

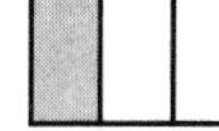 go to page 22.

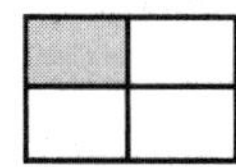 go to page 6.

5

Look out! You step in some quicksand and start to sink! This is not the right answer. Go back to your last page and try again

6

Help! You start to run, but you fall down and twist your ankle! This is not the right answer. Go back to your last page and try again.

7

Yikes! That is one BIG toe! As you and your friend are getting ready to measure the rest of the foot, something grabs you from behind. You turn around to see two huge, purple aliens. Each one has two heads, a tail, and claws. They are 6 feet tall from their feet to their waists, 4 feet tall from their waists to their necks, and 2 more feet from their necks to the top of their heads. How many feet tall is each alien?

If your answer is... 12 feet, go to page 11.
13 feet, go to page 13.
10 feet, go to page 9.
8 feet, go to page 10.
11 feet, go to page 12.

8

E-E-E-K! You get so scared that you run all the way home and never solve the mystery of the alien invaders. This is not the right answer. Go back to your last page and try again.

9

Hiss! You look down. There is a big rattlesnake sitting on your foot, and it looks mad! This is not the right answer. Go back to your last page and try again.

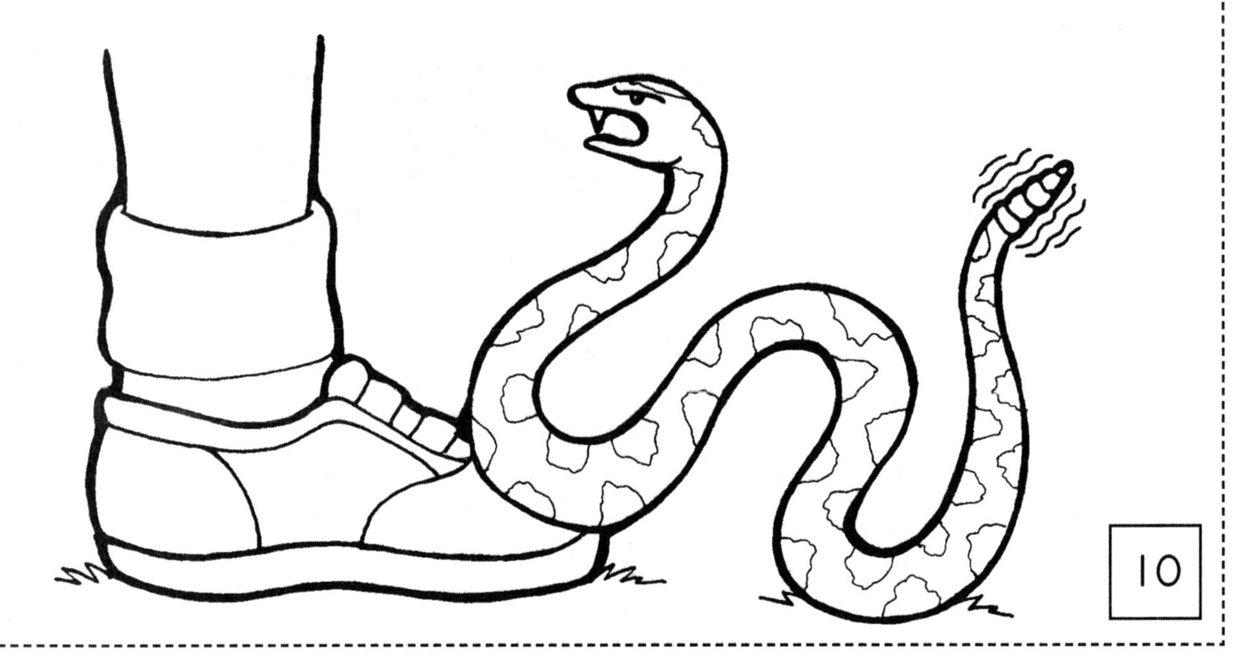

"You are coming with us," hiss the aliens as they drag you off. They stop in front of a big hill and pull on a tree branch. A door swings open, and you are pulled inside. Wow! It's a spaceship!

"Now tell us how we can take over the earth, or we will use our alien magic to make all of the houses in your town vanish!" they say with an evil grin. Oh, no! There are only 20 houses in your small town. If the aliens use their magic on 5 and 3 of them, and then 2 more, how many houses will be left?

If your answer is... 30, go to page 9.

10, go to page 14.

9, go to page 13.

0, go to page 15.

20, go to page 12.

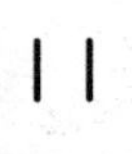

The aliens decide to take you back to their planet and put you in a zoo! This is not the right answer. Go back to your last page and try again.

12

Gasp! You try to run, but the aliens turn you into an alien! This is not the right answer. Go back to your last page and try again.

13

The aliens take you into a room full of flashing buttons. "We will give you a minute to think," the aliens hiss. "Then we will come back. You know what will happen if you do not help us." They thump out of the room. You and your friend look around. You see a button that says "SELF DESTRUCT" high up on a wall. Will that blow up the ship? You have to try it! There are 5 boxes in the room. They are 2 feet tall, 7 feet tall, 6 feet tall, 4 feet tall, and 8 feet tall. The button is about 17 feet up. What 2 boxes can you stack up to be the CLOSEST to 17 feet tall?

If you choose . . 8 and 2, go to page 12.
4 and 6, go to page 13.
7 and 8, go to page 17.
7 and 4, go to page 16.
7 and 6, go to page 15.

14

Zoom! The spaceship takes off with you in it! Will you ever see the earth again? This is not the right answer. Go back to your last page and try again.

15

Yuck! Green slime drips from the ceiling and covers you! This is not the right answer. Go back to your last page and try again.

16

You push the button and a sign lights up. It says, "Ship will explode in 20 seconds! 19 . . . 18 . . . ." You climb down and run to the room's door. It opens, but when you get to the spaceship door, you can't find its handle. You push on the door, but it is stuck. "How much time before we blow up?" your friend yells. It took 3 seconds to climb down the boxes, 5 seconds to run from the other room, and 7 seconds to try to open this door. How much time have you used up?

If your answer is... 15 seconds, go to page 4.
12 seconds, go to page 18.
13 seconds, go to page 21.
14 seconds, go to page 23.
8 seconds, go to page 16.

17

Uh, oh! The ship is shaking and spinning. It's going to explode! Help! This is not the right answer. Go back to your last page and try again.

18

Hooray! You picked the correct clock! The door swings open, and you and your friend dive out the door just as the ship begins to take off. The aliens are trying to get home before the ship blows up! You and your friend see a flash of light. Wow! The ship is gone! You turn around and see a man running up to you. He is a reporter who was looking for the ship. He writes down your story and takes your picture for the newspaper. When you get home, the mayor is so happy to hear that you saved the town, that she throws a big party for you. You are heroes! Great job, Math Detective!

19

Look out! The aliens see you! They save the ship and decide that you are too smart to stay on earth. They take you back to their planet. This is not the right answer. Go back to your last page and try again.

20

Gulp! You hear a funny sound and turn around. Your friend is gone, and there are 10 aliens standing in back of you! This is not the right answer. Go back to your last page and try again.

21

"Wow! Look at the size of those prints!" your friend says. "I bet elephants' feet aren't even that big." "Let's check it out," you reply. You take your ruler out and hold it up to one of the toe prints. Which measurement tells about how many centimeters (cm) long the toe is?

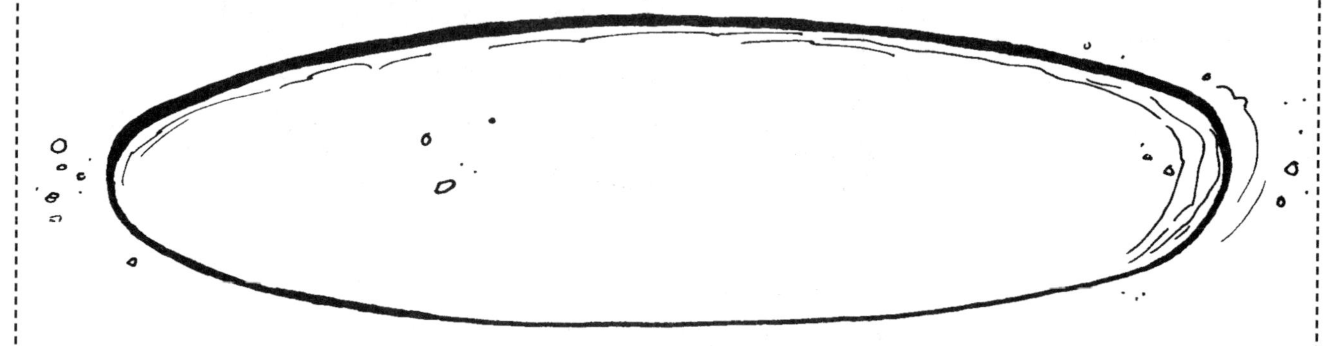

If your answer is...

6 cm, go to page 10.
10 cm, go to page 9.
20 cm, go to page 7.
13 cm, go to page 6.
15 cm, go to page 8.

22

An alien fitness coach walks in the room. "Time to do 1,000 push-ups," it says with an evil grin. This is not the right answer. Go back to your last page and try again.

23

You and your friend have heard that the old, empty house at the end of your street is haunted. You decide to check it out and see what you can find. You tiptoe up to the house and stop. Its front steps are falling apart and do not look safe. The 2 steps at the bottom are okay, and the 3 near the top are good, but the 4 in the middle are rotten. How many steps are safe to walk on altogether?

If you choose...

8 steps, go to page 5.
3 steps, go to page 2.
7 steps, go to page 3.
9 steps, go to page 7.
5 steps, go to page 6.

1

Yikes! A huge snake jumps out from behind a wall and begins moving toward you! This is not the right answer. Go back to your last page and try again.

2

Oh, no! You fall through the floor and hurt your leg! This is not the right answer. Go back to your last page and try again.

3

You can only go in 3 of the rooms. Your friend hears a noise coming from the living room, so you decide to check it out. When you walk into the room, you gasp. It is full of oddly shaped pictures of eyes all staring at you no matter where you move! Look at the pictures below. Which pictures are symmetrical? Use the dashed line as a line of symmetry. (If the shape was folded along the line, the two folded parts would match against each other.)

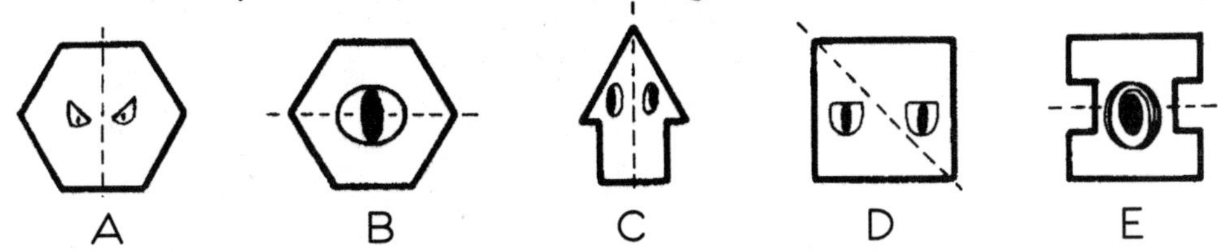

If you choose...

A and B only, go to page 13.
A, B, D, and E, go to page 7.
D and E only, go to page 9.
A, B, C, and D, go to page 11.
All of the shapes are symmetrical, go to page 12.

4

Help! A ghost floats through the wall and carries your friend away. This is not the right answer. Go back to your last page and try again.

5

You make it up the steps and reach the door. It is locked, and there is a big key hole, but you have your trusty set of detective's keys with you. To help you find out if your key will fit, measure the length of the key hole.

If you choose...

1 inch, go to page 7.
4 inches, go to page 3.
2 1/2 inches, go to page 8.
3 1/2 inches, go to page 9.
2 inches, go to page 5.

6

Ouch! A spider runs toward you and is about to bite your big toe! This is not the right answer. Go back to your last page and try again.

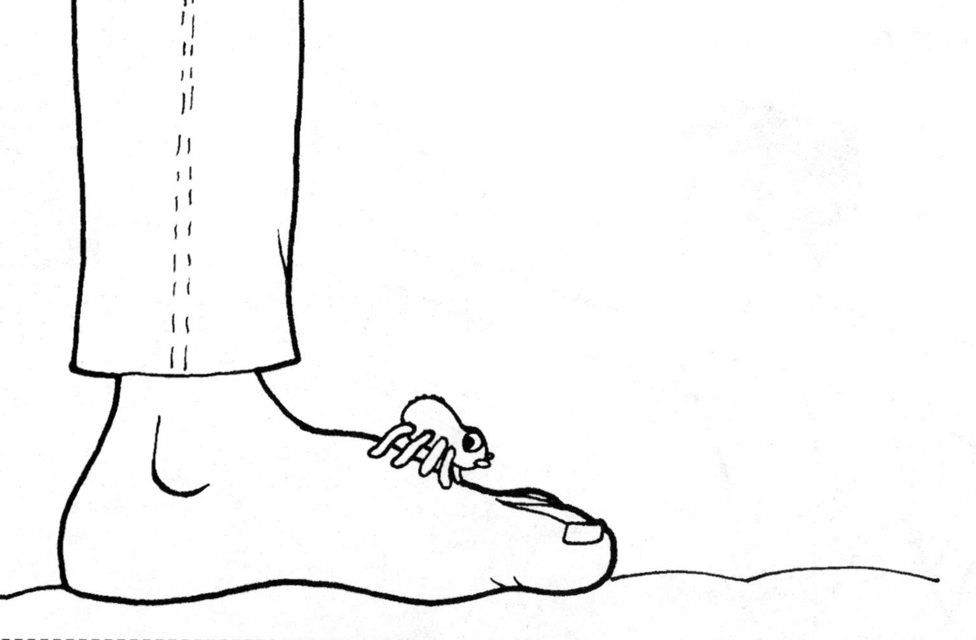

7

Hooray! You find a key the right size, and the front door creaks open. You and your friend walk inside and take a look around. There are 10 rooms in the house altogether. If 4 rooms in the front of the house are nailed shut, and 3 rooms in the back of the house are stuck closed, how many rooms are open for you to check out?

If your answer is... 3 rooms, go to page 4.
7 rooms, go to page 11.
17 rooms, go to page 7.
4 rooms, go to page 5.
14 rooms, go to page 9.

8

Oh, no! You panic and run home! You never solve the mystery of the haunted house. This is not the right answer. Go back to your last page and try again.

9

You push on brick number 5, and the door opens, but there is another door in back of it! You hear the voice say, "If you want to open the last door today, pick both 2/3 fractions. They'll show you the way." (Look at the shaded part of the pictures.)

If your answer is...
A go to page 11.
B go to page 16.
C go to page 13.
D go to page 15.
E go to page 17.

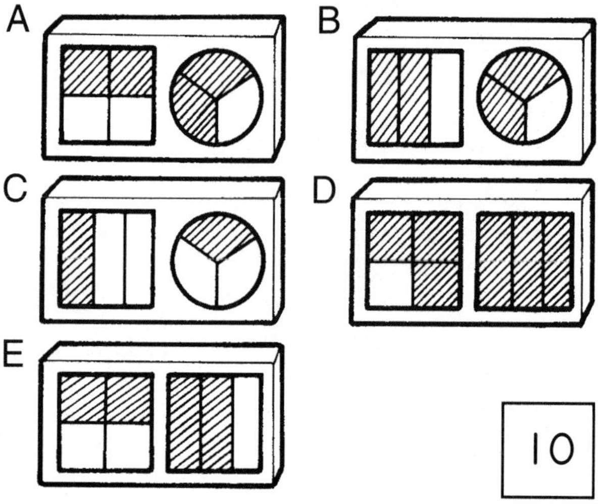

A

B

C

D

E

10

EEK! You become so scared that you can't move, and your hair turns white! This is not the right answer. Go back to your last page and try again.

11

You stare back at all the eyes in the pictures. "BOO!" you yell as loud as you can. Wow! The eyes are gone, but your friend is gone too! You walk over to the fireplace to where your friend had been standing. One of the bricks must have a hidden button behind it to open a trap door. Suddenly, you hear an odd voice say, "Add 2 and 3 to 5 and 4, take away 9, and open the door." Can you solve the problem to find the correct brick?

If your answer...
0, go to page 15.
9, go to page 9.
23, go to page 11.
5, go to page 10.
11, go to page 13.

12

Help! A witch turns you into a frog! This is not the right answer. Go back to your last page and try again.

13

The robot falls over with a crash! You and your friend race out the front door and run all the way home. You call the police and tell them what happened. They arrest the magician and give the gold to the real owners of the house who had been scared away. They are so happy that they give you some of the gold. Now you are rich and a hero. The mayor even makes you the Chief Math Detective of ______________________ (your city)! People come from all over the country to hear your story! Great job, Math Detective!

14

Oh, no! You wake up. It was all a bad dream. Too bad! This is not the right answer. Go back to your last page and try again.

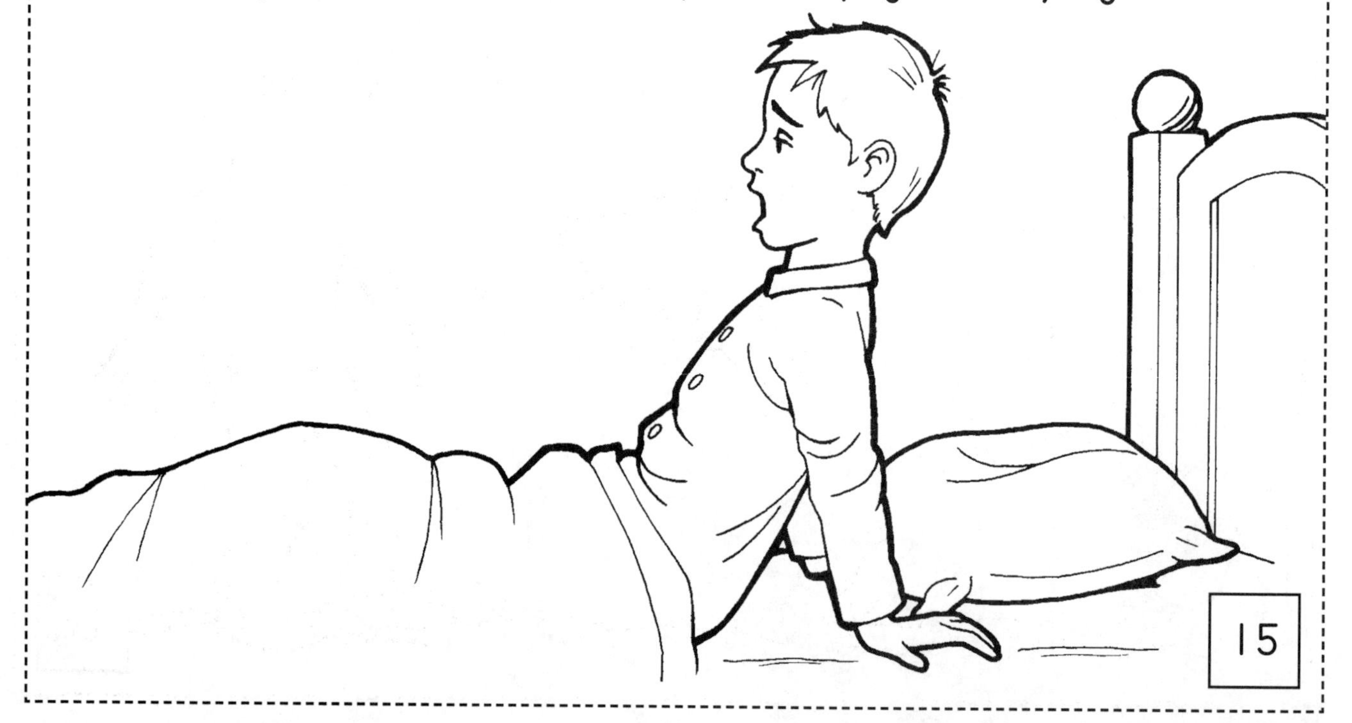

15

Whew! You picked the right fractions. The last door swings open. You step into a tunnel. It is dark and cold. You walk along slowly and almost fall in a big hole full of snarling wolves. There is a bridge hanging over the hole that will come down if you can pull the lever with the correct problem below.

If the problem you choose is...

$10 - 2 + 0 < 2 + 2 + 2$, go to page 17.
$6 + 7 + 1 = 15 - 5 + 1$, go to page 19.
$11 - 6 - 1 > 5 + 6 - 5$, go to page 15.
$12 + 6 - 4 < 7 + 7 + 2$, go to page 21.
$6 + 3 - 1 > 8 - 2 + 4$, go to page 13.

Help! A hungry, growling wolf leaps out right in front of you. This is not the right answer. Go back to your last page and try again.

You push the cube button. The door opens, and you walk into a room filled with computers. Your friend is sitting in a chair next to a magician who is looking at you! You now know that the haunted house is a fake and run by all the computers.

"Yes!" says the magician. "This house is not haunted, but it is sitting on a huge pile of gold. I have to keep scaring people away until I have dug it all up, but now that you know my secret, you will have to stay here forever unless you can answer my riddle!" Then he growls, "If I give you 3 minutes and 10 minutes, take away 8 minutes, then add 4 minutes, how much time will I give you to escape?"

If you choose... 25 minutes, go to page 20.
24 minutes, go to page 22.
8 minutes, go to page 17.
1 minute, go to page 19.
9 minutes, go to page 23.

18

"No!" The magician shouts as he waves his magic wand. This is not the right answer. Go back to your last page and try again.

19

Yikes! A computer bug jumps out and bites you on the ear! This is not the right answer. Go back to your last page and try again.

20

You pull the lever, cross over the bridge, and walk along until you see a locked door. You can hear your friend's voice on the other side! You look at the door. There are some funny shaped buttons sticking out of the wall. "Your friend will never find the cube shape," a voice says from the other side of the door. Can you find the cube button?

If you choose...

go to page 15 go to page 18.

go to page 17. go to page 19. go to page 20.

21

"Wrong!" The magician shouts. "You have to stay here forever!" This is not the right answer. Go back to your last page and try again.

22

"Only 9 minutes!" you yell as you grab your friend and run out of the room. The shocked magician yells, and a robot runs into the room to catch you. It traps you in a corner. You see numbers on the robot's chest as it reaches to grab you. They are 5, 10, 15, __ , 25, __ , __ , 40, 45, __. If you can push the right numbers to finish the pattern, the robot might stop!

If the numbers you push are...

20, 30, 35, and 50, go to page 14.
17, 27, 29, and 47, go to page 17.
24, 26, 39, and 46, go to page 22.
16, 26, 27, and 46, go to page 20.
20, 30, 40, and 50, go to page 19.

23

The Insect Investigation

Detective's Name:

____

Date:

____

DON'T GO BUGGY!

You and your friend have been sent to capture the evil Dr. Bug-Eyes who wants to take over the world and turn everyone into insects. You walk up to the doctor's lab. It looks like a huge beehive. You do not see a door, but you do see a hole that is shaped like a cone or a large bee stinger. You must pick the correct shape below to fit in the hole.

If you choose...

A go to page 9.
B go to page 6.
C go to page 2.
D go to page 3.
E go to page 5.

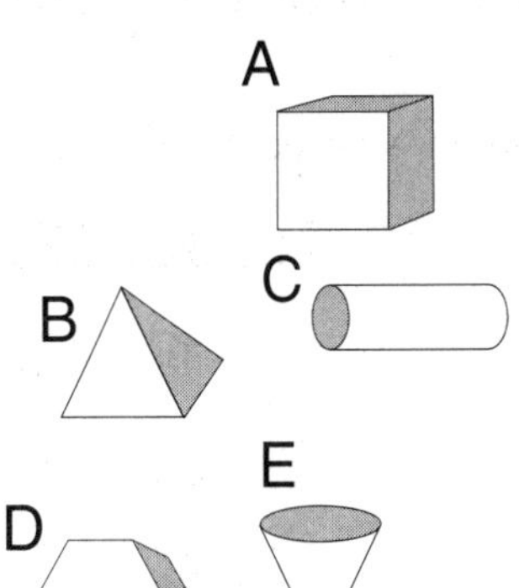

1

Help! A huge swarm of angry bees is flying towards you! This is not the right answer. Go back to your last page and try again.

2

Oh, no! Ants come pouring out of a nearby nest and cover you! This is not the right answer. Go back to your last page and try again.

3

"Nice job," the spider says as he sets you free. "I also don't like that doctor. She tried to get rid of me because I was eating some of her insect friends. I'll help you get her. Finish the pattern on the wall and one of the hexagons will open. The doctor is hiding in a room behind it."

You and your friend thank the spider and hop up the wall to the numbers. They are 21, 19, 17, ___, ___, 11, ___, 7, ___, 3. Can you find the numbers to finish the pattern?

If you choose...

18, 19, 12, and 8, go to page 18.

13, 12, 8, and 4, go to page 12.

16, 15, 8, and 4, go to page 15.

16, 15, 10, and 6, go to page 16.

15, 13, 9, and 5, go to page 14.

4

A secret door opens in the beehive wall. You and your friend walk inside. Help! All of a sudden you can't move. Your feet are covered in sticky honey. You grab a nearby cup and spoon and start scooping the honey into the cup with the spoon. The cup holds 4 spoonfuls of honey, but your feet are covered in 24 spoonfuls of honey. How many cups must you fill before you are free?

If your answer is...

6 cups, go to page 7.

28 cups, go to page 3.

7 cups, go to page 6.

20 cups, go to page 2.

5 cups, go to page 9.

5

Yikes! A big mosquito is flying towards you, and he looks hungry! This is not the right answer. Go back to your last page and try again.

6

You finally scoop up the last of the honey. You and your friend walk into a room. It looks like a huge honeycomb because the walls are covered with hexagon-shaped openings just like in real hives! How many sides does a hexagon have? How many sides would three hexagons have altogether? You must answer both questions to find out what is hiding in this room!

If your answer is... 4 and 12, go to page 6.
8 and 24, go to page 12.
5 and 15, go to page 10.
6 and 12, go to page 9.
6 and 18, go to page 11.

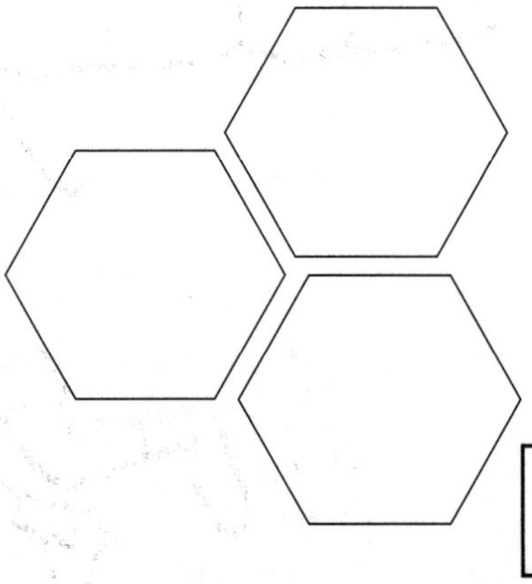

7

What? How can you be that tiny? You look around and see that the doctor has escaped. Then you look at your friend. Oh, no! Your friend now has 3 body parts, 6 legs, 2 antennae, 5 eyes, and 4 wings. How many buggy parts does your friend have in all ?

If you choose... 11, go to page 15.
21, go to page 12.
20, go to page 13.
9, go to page 10.
19, go to page 16.

8

Help! A huge dragonfly swoops down and carries you away! This is not the right answer. Go back to your last page and try again.

9

Oh, no! A caterpillar carries you off and wraps you up in a cocoon! This is not the right answer. Go back to your last page and try again.

10

The three hexagons that you were looking at swing open. It's Dr. Bug–Eyes. She was hiding in the room! You run to grab her, but she pulls out a little metal box. "Insects rule!" she screams as she pushes the button on the box. You and your friend start to shrink! If you can lift your big detective's kit and get your ruler, measure how tall you are now in centimeters (cm) and inches. Your new height is shown below.

If your answer is... 5 cm, 5 inches, go to page 10.
1 cm, 2 inches, go to page 9.
2.5 cm, 1 inch, go to page 8.
3 cm, 1 inch, go to page 12.
1 cm, 2 inches, go to page 15.

11

EEK! A big snake slithers toward you. Lunch time! This is not the right answer. Go back to your last page and try again.

12

Yikes! Your friend has turned into a strange–looking bug... and you have, too! You both start hopping around in shock when all of a sudden you can't move. You are stuck on a spider's web! "Hello there," says a big, black spider. Then it says, "20 loses 15 but adds 5 and 3. Answer my riddle and I'll set you free." Can you do it?

If your answer is... 13, go to page 4.
23, go to page 12.
28, go to page 2.
3, go to page 6.

13

You got the pattern! The wall swings open, and you hop inside. You see Dr. Bug-Eyes talking to a big hornet. "The math detectives will stay insects forever if they can't catch me by 2:30." You look at the clock on the wall. What time is it now, and how much time do you have left before it turns 2:30? You must solve both problems.

If your answers are... 2:00 and 10 minutes left, go to page 16.
2:15 and 10 minutes left, go to page 22.
3:10 and 15 minutes left, go to page 18.
2:15 and 15 minutes left, go to page 19.
3:10 and 20 minutes left, go to page 20.

14

Help! A stinkbug jumps onto you! Oh, yuck! This is not the right answer. Go back to your last page and try again.

15

Oh, no! A big fly swatter is swinging right at you! This is not the right answer. Go back to your last page and try again.

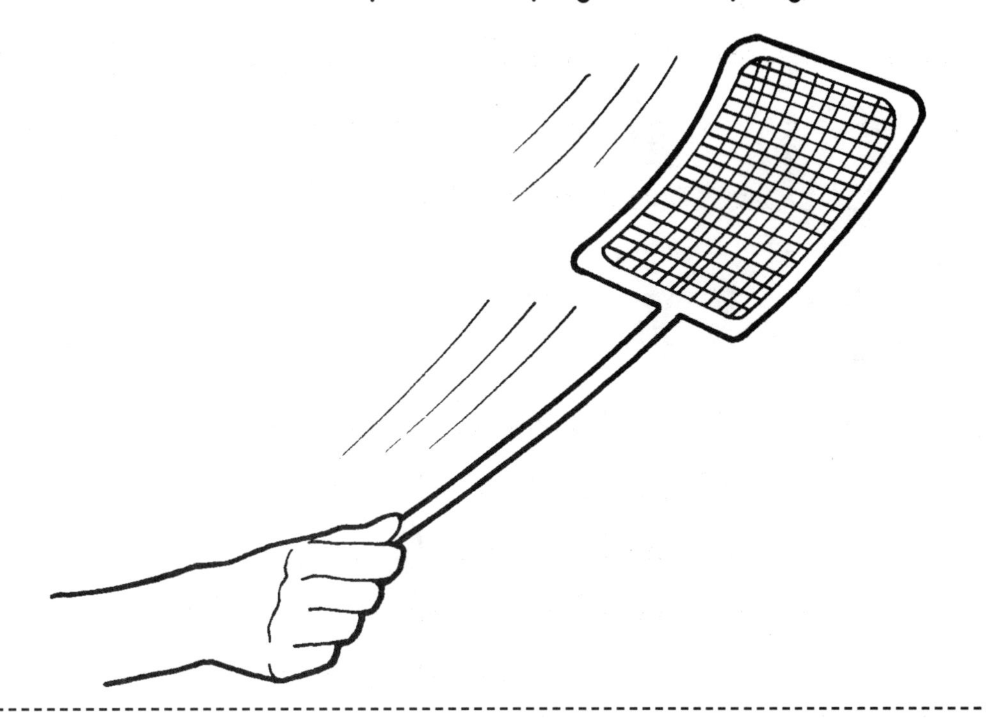

16

You picked the right fractions. Then you notice that a spray can is sticking out of the doctor's pocket. It says "People Spray" on it. It must be the spray that will turn you back again...if you get to it in time. You had 15 minutes left to try to turn back to people, but you used up 4 minutes looking around the room and then you used up 8 minutes thinking up a plan. How many minutes do you have left before you stay insects forever?

If you answer...
3 minutes, go to page 23.
11 minutes, go to page 22.
4 minutes, go to page 18.
27 minutes, go to page 20.
12 minutes, go to page 16.

17

Help! A big, hungry frog is hopping right at you! This is not the right answer. Go back to your last page and try again.

18

Only 15 minutes until you stay an insect forever! You look around the room. On the wall are some charts showing how many people in some of the nearby towns have been turned into insects. Which fractions below match the charts? Be fast! Your town is next! (Look at the shaded part of the pictures.)

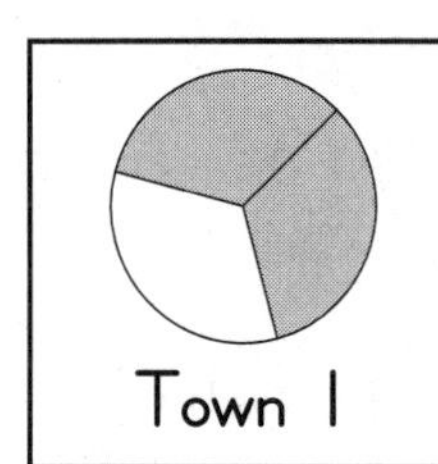
Town 1

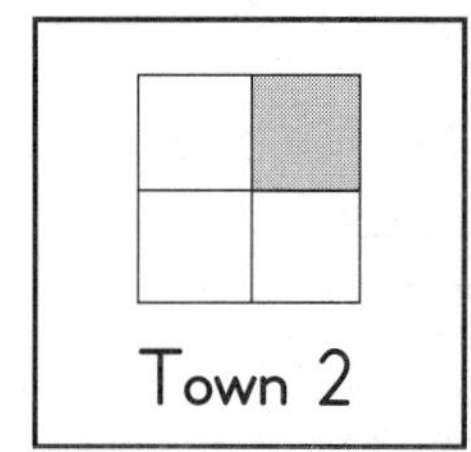
Town 2

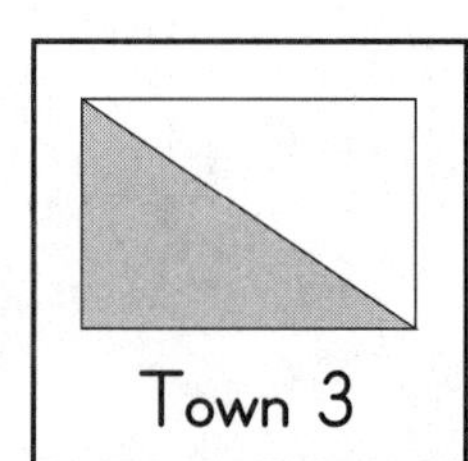
Town 3

If you choose...

1/3, 1/4, and 1/2, go to page 15.
2/3, 1/4, and 1/2, go to page 17.
2/3, 1/3, and 3/4, go to page 18.
1/2, 1/4, and 1/2, go to page 16.
3/4, 1/4, and 1/3, go to page 20.

19

"You'll never get away from me!" shouts the doctor as she starts spraying her magic fog. When the fog clears, you see that the room is full of huge insects ready to jump on you and cover you with bug stickers and stamps. This is not the right answer. Go back to your last page and try again.

20

You push the shapes. The room fills up with fog, and you and your friend become people again. That's too bad for the doctor. You were in her pocket, and now you are sitting on top of her. She can't move! You use one of her sprays to turn her into a harmless butterfly, and she flies away. You call the police, and they help you turn everyone back into people. You have saved the world again! You become so famous that you are made Chief Math Detective of the Nation! Great job, Math Detective!

21

"Wrong!" Doctor Bug-Eyes shouts. "Your time is up! Now you will be insects forever! Ha! Ha! Ha!" This is not the right answer. Go back to your last page and try again.

Only 3 minutes left! You and your friend begin making cricket-sounds and soon the doctor gets sleepy and her eyes start to close. Quickly you hop into her pocket and land on the can. It has buttons of different shapes written on it with a code that says, "Push the sphere; then the hexagon, the pyramid, and the cylinder." Pick the right buttons quickly. Your time is up!

If you choose...

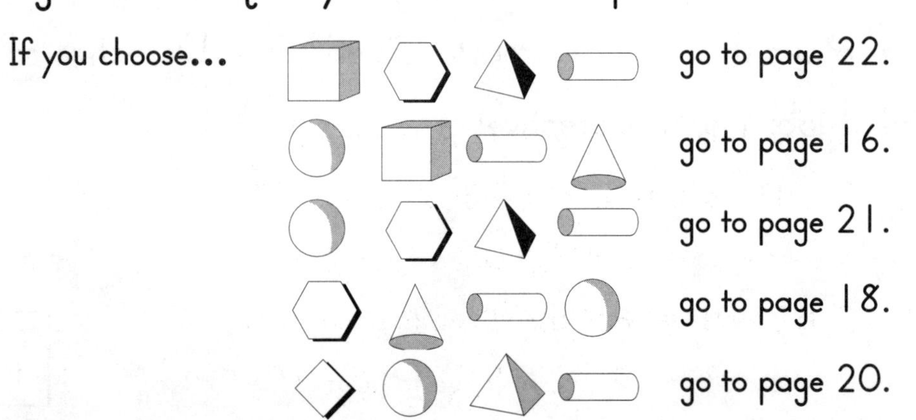

go to page 22.

go to page 16.

go to page 21.

go to page 18.

go to page 20.

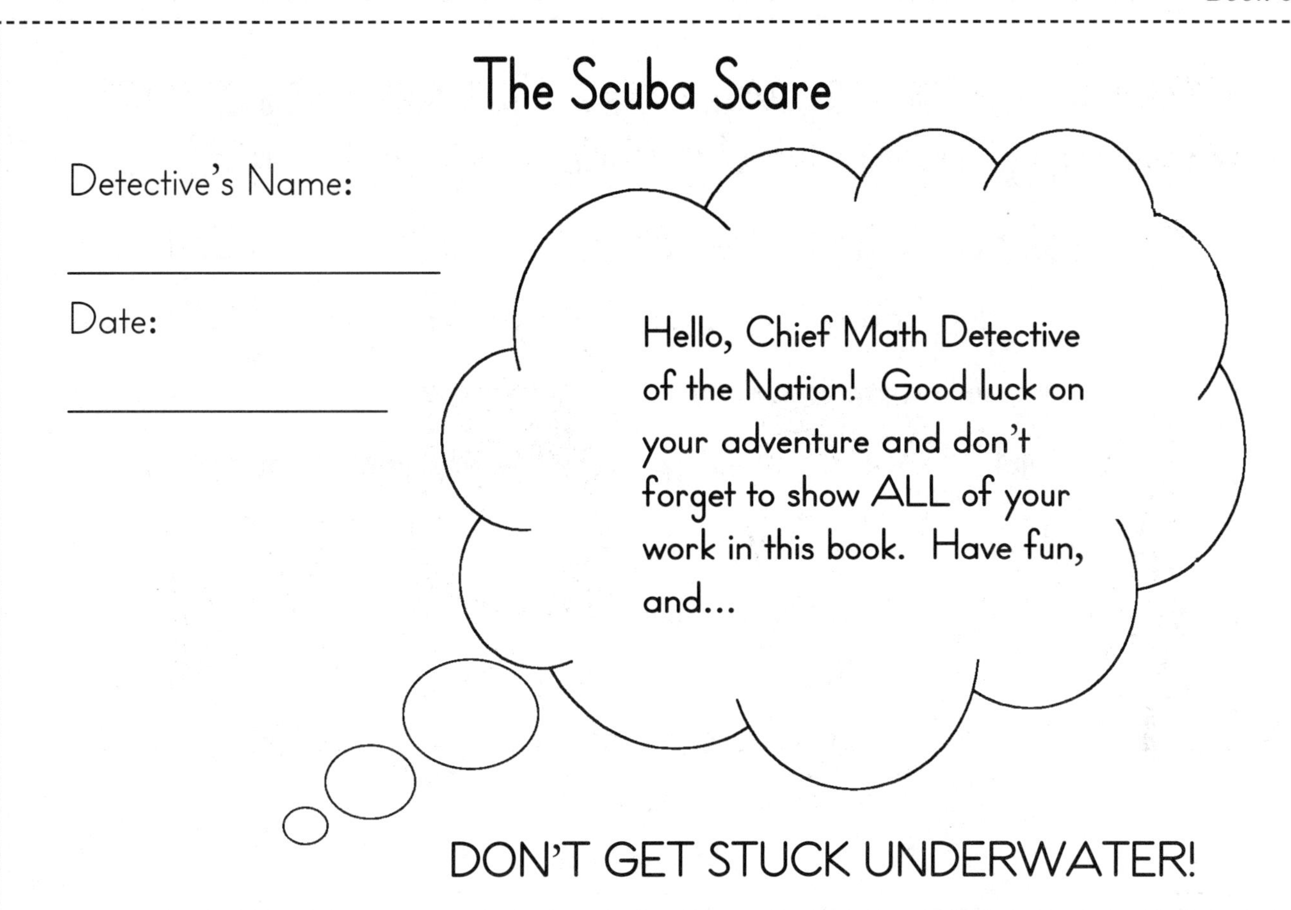

You and your friend are tired and need a vacation from all of your detective work, so you decide to go scuba diving for the day. You climb onto the boat. There are several buttons, but none of them say "go." You must choose the correct problem below in order to find the button that starts the boat.

If you choose...

$6 - 4 + 2 > 5 + 2 + 1$, go to page 7.

$0 + 10 - 2 = 6 + 4 - 3$, go to page 12.

$8 - 4 - 2 > 3 + 3 + 0$, go to page 15.

$2 + 2 - 4 = 6 - 5 - 1$, go to page 9.

$7 - 4 + 3 < 8 - 2 - 5$, go to page 19.

You pick up the shell and put it to your ear to listen to the sea. What? The shell is talking! It says, "Finish my pattern. If this you do, I will grant one wish for you. It is 27, 24, 21, 18, 15, 12, ___, ___, ___." Before you can answer you hear a CRACK! The octopuses are pulling the boat apart. "Help! I wish we were home!" you yell. Quick! Finish the shell's pattern before your boat sinks!

If you choose...

11, 10, and 9, go to page 22.
10, 8, and 6, go to page 12.
8, 4, and 0, go to page 15.
9, 6, and 3, go to page 10.
13, 14, and 15, go to page 19.

2

Oh, no! You get trapped in a school of fish. Now you will have to stay underwater forever. This is not the right answer. Go back to your last page and try again.

3

You pick up the coins and go deeper inside the cave. You turn on your flashlight and see an animal stuck in a pentagon-shaped fishing net just ahead. If the net is 4 meters long on one side, 2 meters long on two sides, and 3 meters long on two more sides, what is the perimeter of the net (or how many meters around is it altogether), and how many sides does a pentagon have?

If your answers are... 14 meters and 5 sides, go to page 8.
9 meters and 3 sides, go to page 3.
14 meters and 6 sides, go to page 17.
9 meters and 5 sides, go to page 5.
11 meters and 4 sides, go to page 11.

4

Help! You feel dizzy! Your air is all used up! This is not the right answer. Go back to your last page and try again.

5

Help! Your air tank's breathing tube has sprung a leak! This is not the right answer. Go back to your last page and try again.

6

Hey! The bottom of the boat falls off, and you drop into the VERY cold water. This is not the right answer. Go back to your last page and try again.

7

As you swim up to the net, you see that it is not an animal in the net. It is your friend! You begin tugging at the net when three big octopuses swim up to you waving their arms so they can grab you. How many arms will you have to get away from in all?

If your answer is... 3 arms, go to page 20.
18 arms, go to page 5.
6 arms, go to page 17.
32 arms, go to page 3.
24 arms, go to page 18.

8

You push the correct button, and the boat starts up. As you drive, you and your friend check your watches. It is 7:50 A.M. now. You are due to arrive at your scuba diving spot at 10:17 A.M. Choose the clocks below that picture those two times.

If you choose...

 go to page 19.

 go to page 7.

 and go to page 23.

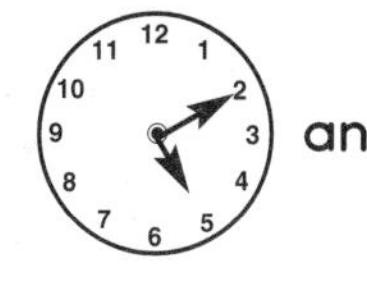 go to page 12.

and go to page 15.

9

When the shell hears your numbers, it vanishes. Zing! You and your friend are safe at home. "Wow! Some vacation!" says your friend.

When the local TV station hears about your thrilling trip, you end up on TV all over the world. You are famous in every country, so you become Chief Math Detective of the Whole World! Great job, Math Detective!

10

Ouch! You swim into a jellyfish, and it stings you! This is not the right answer. Go back to your last page and try again.

11

Help! A huge wave covers the boat and washes you overboard. This is not the right answer. Go back to your last page and try again.

12

You and your friend swim out of the mouth before the whale even knows what happened. You begin to swim to the boat when you start to feel funny. You check your air tank and see that 4/5 of your air is gone. Find the picture below that shows 4/5 and the fraction that shows how much air you have left. Hurry! Here come the octopuses! (Look at the shaded part of the pictures.)

If you choose...

picture A, go to page 14.
picture B, go to page 5.
picture C, go to page 11.
pictures B and C, go to page 20.
pictures A, B and C, go to page 17.

A

B

C

13

You and your friend swim up to your boat just as the octopuses reach you and you run out of air. You climb onto the boat gasping. As you rest, your friend says, "Look what I found inside the whale before I got trapped in the net." Your friend hands you a symmetrical (if it could be folded along the dashed line, the two halves would match) shell. Which of the pictures below is the shell?

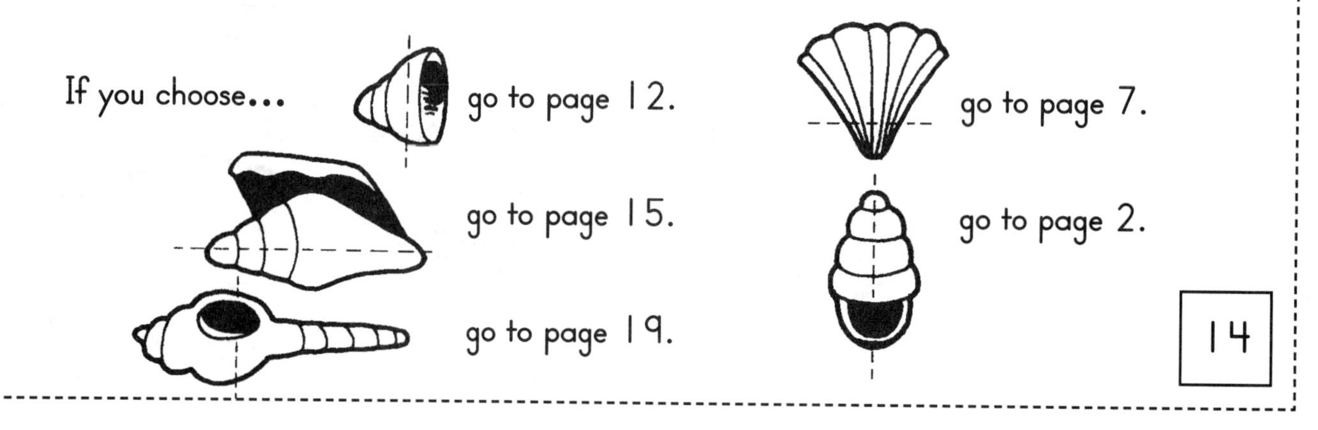

Gulp! A swordfish cuts a hole in your boat, and it sinks. This is not the right answer. Go back to your last page and try again.

The shark finally swims away when you notice that your friend is gone! You decide to swim back to the boat to see if your friend is there. On your way back, you see a cave that you missed before. As you peek inside, you see some coins scattered around on the floor. You pick up 1 quarter and 10 pennies. As you enter the cave, you find 2 nickels and 1 dime. How much money do you have altogether?

If you have...

60¢, go to page 17.
50¢, go to page 5.
55¢, go to page 4.
14¢, go to page 3.
75¢, go to page 20.

16

Yikes! A group of hungry sharks has mistaken you for a yummy seal. Lunch time! This is not the right answer. Go back to your last page and try again.

17

You give the net a big yank and free your friend. You throw the net over the octopuses and swim like mad to the opening of the cave. Yikes! It is blocked by a row of teeth! You are inside a whale's mouth! Luckily, it is missing some teeth. You just have to find a hole big enough to swim through. If you are 18 inches wide with your air tanks on, and each tooth is 3 inches wide, at least how many teeth need to be missing in a row for you and your friend to get out of the mouth?

If your answer is... 21 missing teeth, go to page 20.
5 missing teeth, go to page 5.
15 missing teeth, go to page 17.
6 missing teeth, go to page 13.
7 missing teeth, go to page 11.

18

Ahhhhh! Smoke is pouring from your boat. The motor is on fire! This is not the right answer. Go back to your last page and try again.

19

Oh, no! An electric eel swims up and wraps itself around you. Help! This is not the kind of hug you like! This is not the right answer. Go back to your last page and try again.

20

You reach the sea floor and stop to rest and watch some fish, when suddenly a shark swims up. You dive behind a big rock to hide. The shark chomps down 4 fish, then two groups of 3 fish, and begins to grab 7 more fish, but 2 get away. It swallows the rest. How many fish did the shark end up eating in all?

If you answer... 16, go to page 20.
19, go to page 11.
18, go to page 17.
15, go to page 16.
12, go to page 15.

21

"No!" the shell yells. It vanishes, and so does your boat. You are stuck in the water with the angry octopuses! This is not the right answer. Go back to your last page and try again.

At 10:17 you arrive at the diving spot, turn off the boat, and lower the anchor. You put on your scuba suits and air tanks and jump off the boat. Swimming with tanks is hard work. You have to stop and rest every 5 minutes. If you swim for a total of 30 minutes, how many times will you rest altogether between when you start and when you finish?

If you answer... 5 times, go to page 21.
6 times, go to page 17.
25 times, go to page 5.
35 times, go to page 20.
7 times, go to page 3.

The Future Flight

Detective's Name:

Date:

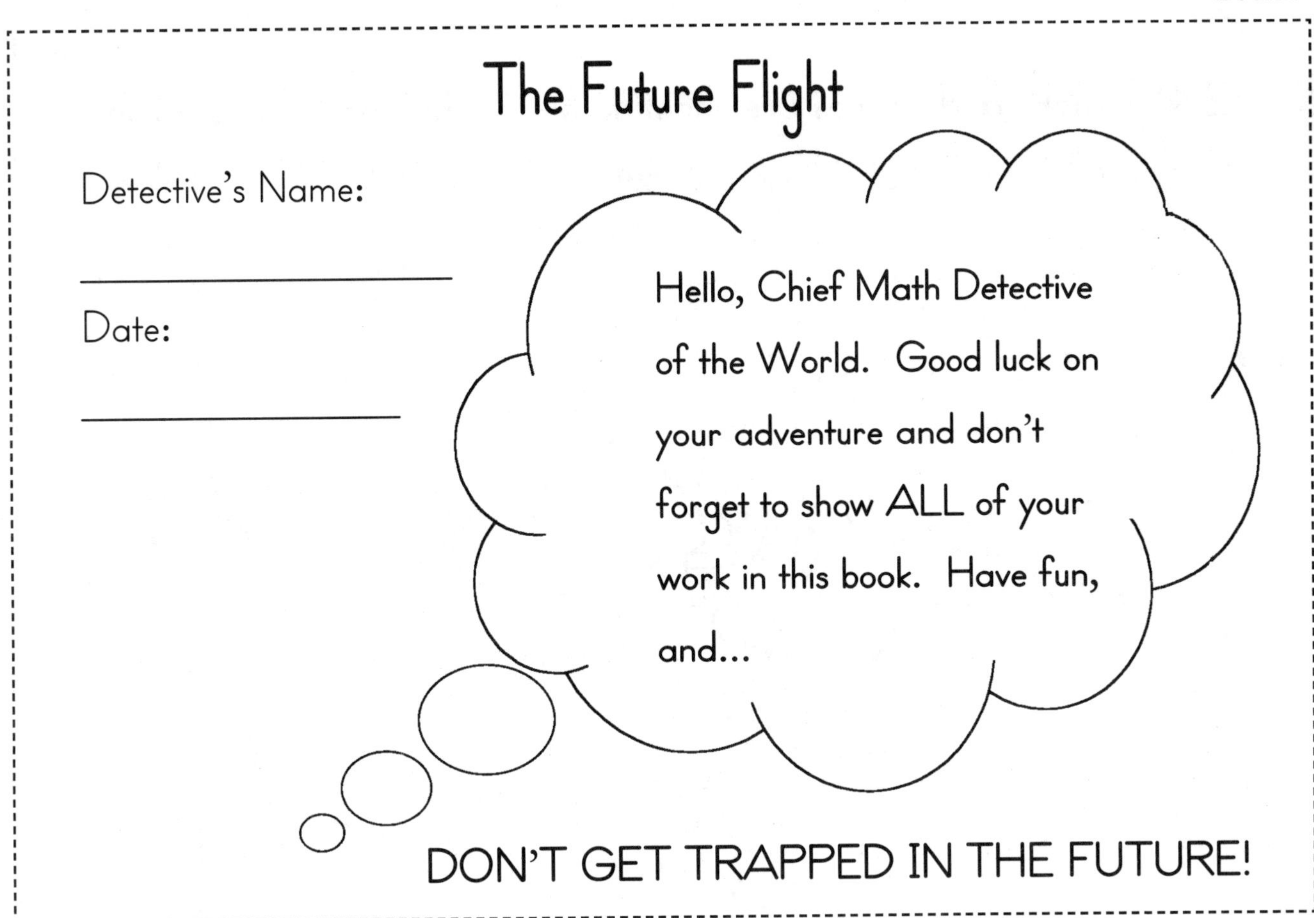

You and your friend have been asked to check out a video arcade where 5 people have vanished. You walk around the video games. One of the newer ones, called Future Flight, looks like fun, so you decide to try it. You sit down to play it like a driving game, but it has buttons like a spaceship might have. You do not even put tokens or money into it to make it start. You have to pick the button that has the correct problem about money on it to begin the game. Can you do it?

If you choose...

1 quarter + 2 pennies > 3 dimes, go to page 2.

2 nickels – 5 pennies = 1 dime – 2 nickels, go to page 4.

1 dime > 1 quarter – 2 dimes, go to page 6.

4 nickels < 1 quarter – 1 nickel, go to page 22.

2 dimes + 3 pennies = 1 quarter, go to page 21.

1

EEK! Smoke and flames are shooting out of the game. It is going to explode! This is not the right answer. Go back to your last page and try again.

2

"No! Fake money!" the android yells. You check your pockets for anything else. You pull out some video game tokens. "Ooh! Pretty!" the android says as it hands you the jet packs. Then it says "police station" as a TV screen rises from its head showing a tall, cylinder–shaped building. Which shape below could be the cylinder–shaped police station, and is it symmetrical or not?

If you choose...

and symmetrical, go to page 17.

and not symmetrical, go to page 12.

and not symmetrical, go to page 21.

and symmetrical, go to page 4.

and symmetrical, go to page 10.

3

You leave the store and try on the jet packs. Maybe they can fly you to the police station. Each pack has a small card hanging on it showing an eye with a glowing pupil in the center. You decide to study it. You take out your ruler to measure how wide the pupil is. You want an accurate measurement. Would inches, centimeters (cm), or millimeters (mm) be the best measurement to use? And, what is your measurement?

If your answers are...

cm, 4 cm, go to page 21.

mm, 4 mm, go to page 18.

inches, 1 inch, go to page 12.

mm, 1 cm, go to page 10.

mm, 14 mm, go to page 17.

4

Gulp! Your jetpack suddenly stops, and you are VERY high up in the sky with no parachute! This is not the right answer. Go back to your last page and try again.

08

5

As you push the button, the game begins to shake, and the words "Can you survive a visit to the year 2501?" appear on the video screen. The room starts to blur around you, and you are no longer at the arcade. You are rocking back and forth on top of a big sphere balanced on a pyramid-shaped building! Which shapes below could be what you are rocking on top of?

If you choose...

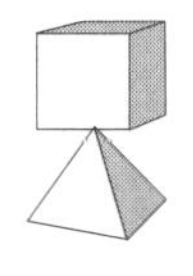 go to page 7.

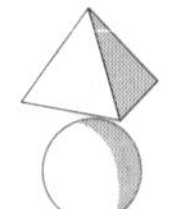 go to page 2.

 go to page 10.

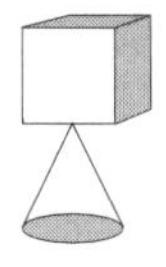 go to page 14.

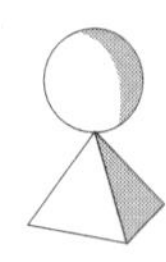 go to page 11.

6

Oh no! The video screen flashes the word "Goodbye!" and takes off. You are stuck in the future. This is not the right answer. Go back to your last page and try again.

7

Help! A pack of flying, android dogs is after you. This is not the right answer. Go back to your last page and try again.

8

You step out of the box. Your friend says, "We'd better find that game or we'll be stuck here forever!" You walk into a store. An android clerk walks up to you. "Where is the police station?" you ask. "Buy first! Questions last!" the android says. You and your friend grab the closest things to you...some funny balloons with straps. "Two jet packs at $10.00 each," says the android. You hand it a $50.00 bill. How much change will you get back?

If your change is... $65.00, go to page 10.
$35.00, go to page 8.
$30.00, go to page 3.
$20.00, go to page 12.
$10.00, go to page 21.

9

Yikes! Some scientists of the future catch you and are going to examine you to find out more about the people who lived in the past. This is not the right answer. Go back to your last page and try again.

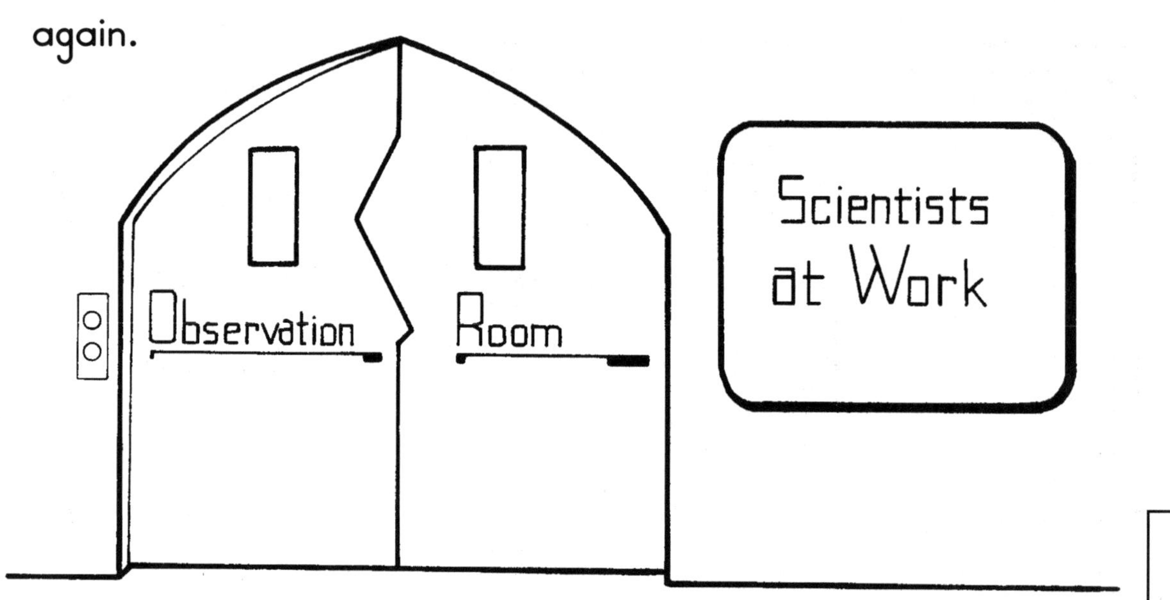

10

Just as you think you might fall off the top of the pyramid building, a tow truck flies up to you and talks! "You may not park here! I must take your car to the police station," it beeps. Before you can talk, it flies away with the video game! If a flying tow truck earns 10 zerks (money of the future) a day, how much would it earn altogether in 1 week?

If you answer... 17 zerks, go to page 8.
14 zerks, go to page 10.
60 zerks, go to page 21.
70 zerks, go to page 16.
3 zerks, go to page 12.

11

Hey! A zookeeper from the future grabs you, puts you in a cage, and hangs up a sign that says "Kid from the Past!" This is not the right answer. Go back to your last page and try again.

12

One of the girls you are saving has a hair clip that size to plug the leak. You push the start button again. The game shakes and sparks fly out of it. You hope 7 people will not be too heavy for it! Suddenly everything blurs, and you are back in the arcade. You check the clock and calendar on the wall. They show that it is 2 days and 10 minutes earlier than when you left! If it was Tuesday at 12:25 when you left, what day and time is it now?

If your answer is... Sunday at 12:15, go to page 15.
Monday at 12: 05, go to page 2.
Wednesday at 2:25, go to page 22.
Thursday at 12:35, go to page 21.
Saturday at 10:25, go to page 14.

13

Oh, no! The video game shakes. It takes you back to the past...the year 1899! Yikes! This is not the right answer. Go back to your last page and try again.

14

"Wow! It's only Sunday!" your friend says just as flames begin to shoot from the game. "Run!" you yell as you lead everyone out of the arcade to the parking lot across the street. A loud "BOOM!" knocks you off your feet. All that is left of the arcade is a big hole in the ground.

When the 5 people tell everyone about their rescue, you are made Chief Math Detective of the Whole Universe, and you are asked to make your adventures into books for school children so that they can become math detectives, too. Great job, Math Detective!

15

As the tow truck flies off, a big box floats up to you. You think that it might be an elevator, so you step into it. There is 1 button on the inside. Every time you push it, the box drops down 9 feet. If you are 27 feet above the ground, how many times will you have to push the button before you can reach the ground?

If you push the button... 36 times, go to page 21.
3 times, go to page 9.
37 times, go to page 12.
18 times, go to page 8.
4 times, go to page 10.

16

Help! The police lock you up for wearing outdated clothing and disturbing the peace. This is not the right answer. Go back to your last page and try again.

17

As you look at the eye, you think of the police station. Suddenly the eyeball begins to flash, and you are lifted up in the air. "Our thoughts control it!" you yell to your friend as you zoom off. It is a rough ride. If you fly forward 4 kilometers, back 2 kilometers, forward 3 more kilometers, and then back 1 kilometer, how far forward have you gone altogether?

If you choose... 7 kilometers, go to page 19.
3 kilometers, go to page 8.
4 kilometers, go to page 23.
10 kilometers, go to page 5.
9 kilometers, go to page 12.

18

Groan! You are stuck in a huge, flying car traffic jam! Some things never change! This is not the right answer. Go back to your last page and try again.

19

You, your friend, and the 5 people pile onto the video game. You push the correct starting button, but nothing happens. Then you notice an odd-shaped hole near the bottom. Fuel is leaking from it. You must use your ruler to find the perimeter of the hole in centimeters so you will know what size plug you will need to stop the leak.

If you choose... 13 cm, go to page 2.
14 cm, go to page 13.
15 cm, go to page 12.
12 cm, go to page 7.
16 cm, go to page 14.

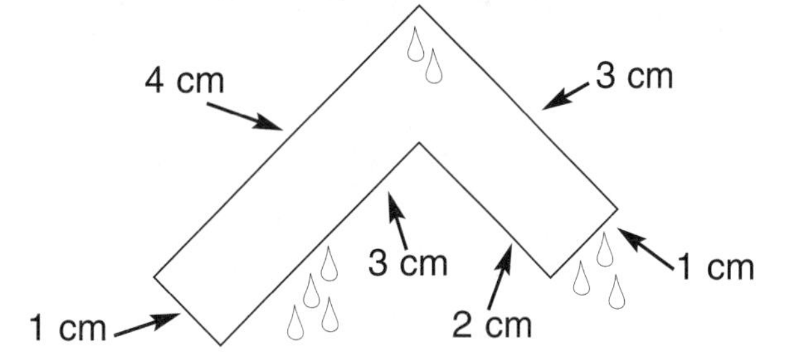

20

Help! A tiny computerized robot flies into your ear and begins draining your brain of everything but math facts. Soon you won't even remember who you are! This is not the right answer. Go back to your last page and try again.

21

Ow! Ow! All of the video games in the arcade are going crazy and shooting tokens at you as fast as they can. This is not the right answer. Go back to your last page and try again.

22

You land at the police station and go inside. “Oh, no! Not more of you!” an officer groans as she points to a bench. The 5 missing people from the arcade are sitting there! They are very happy to see you! The police are so glad that you are handling the case that they give you back the video game and offer you a fraction of space fuel for the trip home. Which one will give you the most fuel? (Look at the shaded part of the pictures.)

If you choose...

2/3, go to page 20.

1/3, go to page 19.

1/4, go to page 17.

2/5, go to page 2.

1/2 (or 2/4), go to page 7.

23

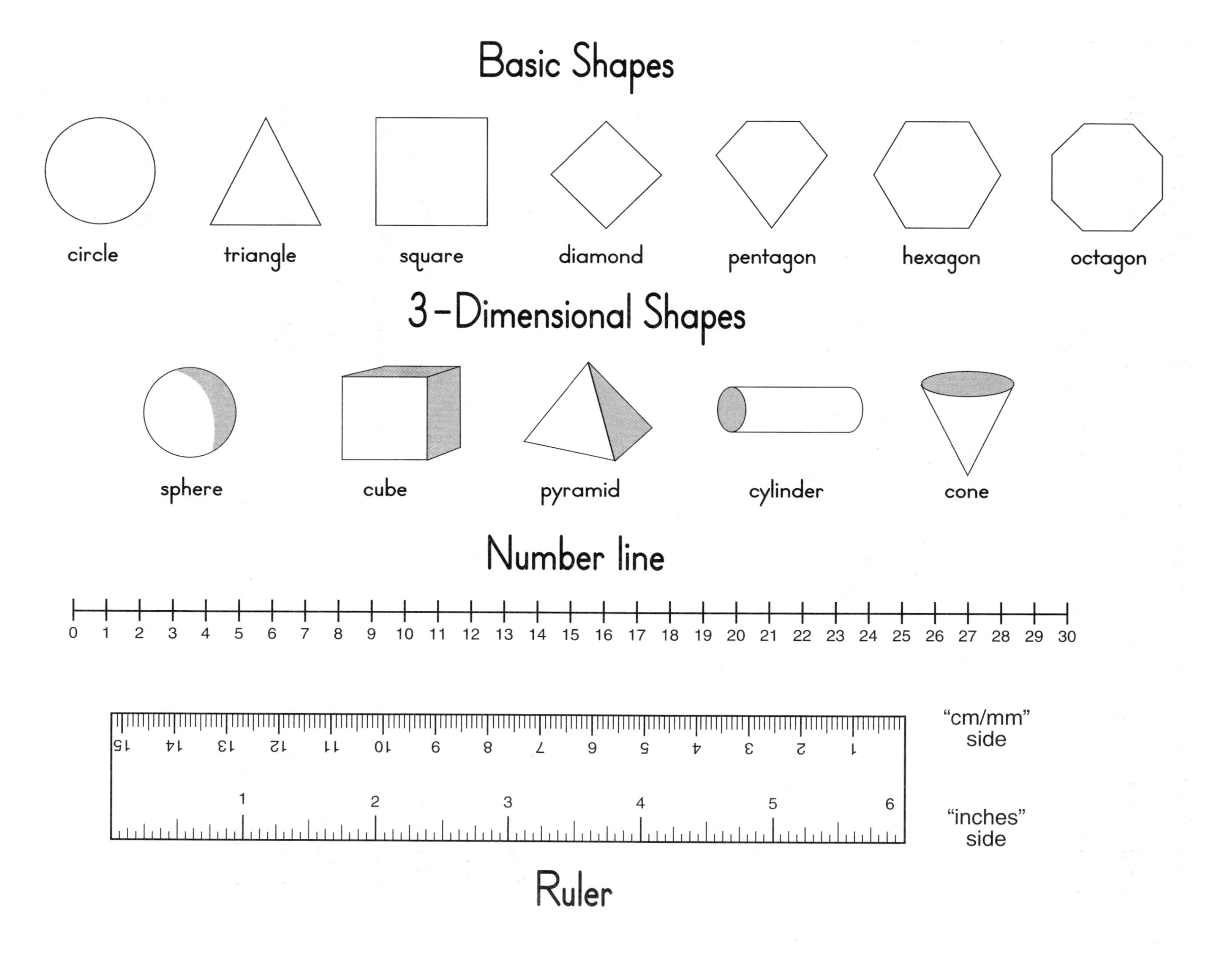
Basic Shapes
circle
triangle
square
diamond
pentagon
hexagon
octagon
3-Dimensional Shapes
sphere
cube
pyramid
cylinder
cone
Number line
0 1 2 3 4 5 6 7 8 9 10 11 12 13 14 15 16 17 18 19 20 21 22 23 24 25 26 27 28 29 30
1 2 3 4 5 6 7 8 9 10 11 12 13 14 15
"cm/mm" side
1 2 3 4 5 6
"inches" side
Ruler

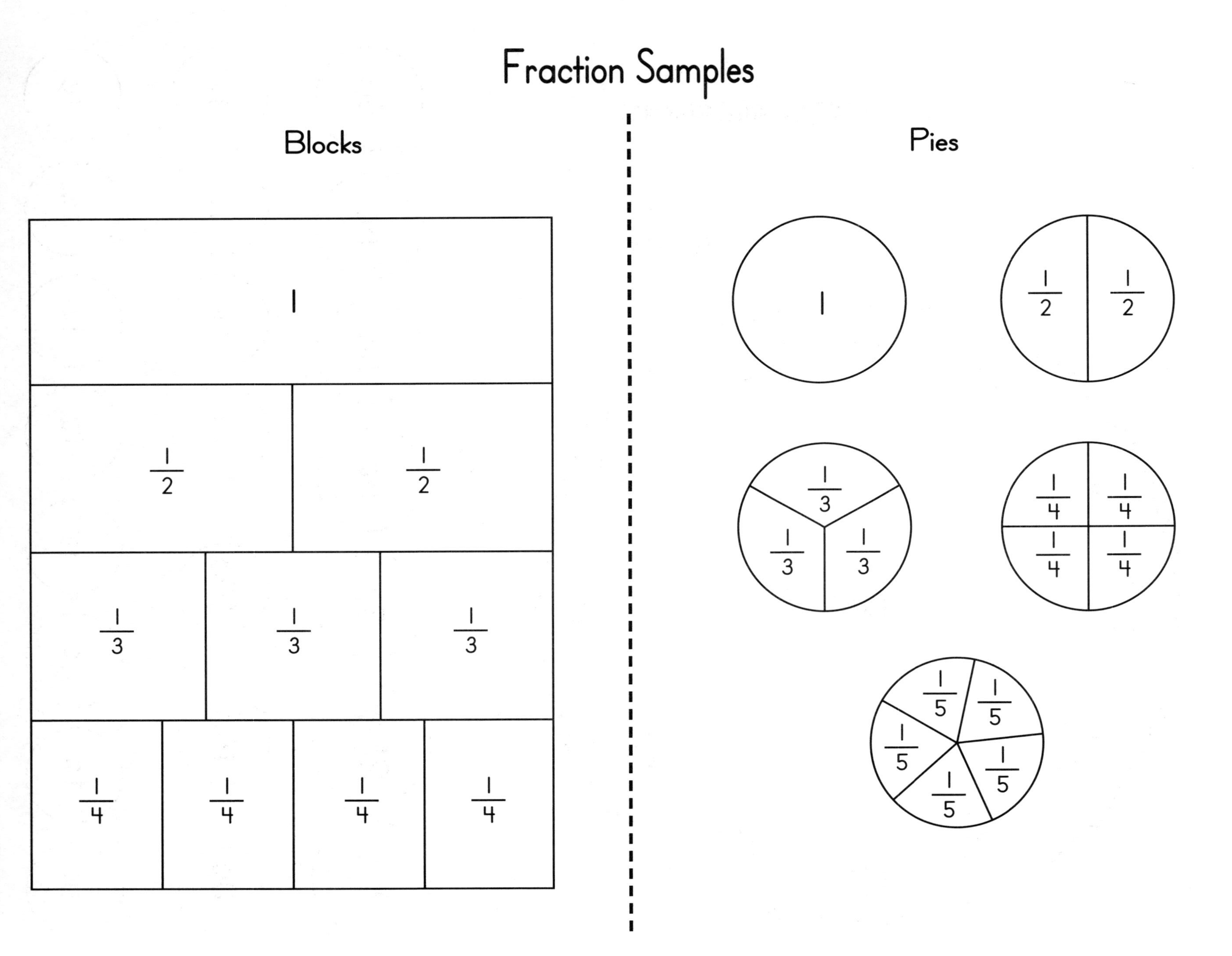
Fraction Samples
Blocks
1
1/2
1/2
1/3
1/3
1/3
1/4
1/4
1/4
1/4
Pies
1
1/2
1/2
1/3
1/3
1/3
1/4
1/4
1/4
1/4
1/5
1/5
1/5
1/5
1/5

Clock Pattern

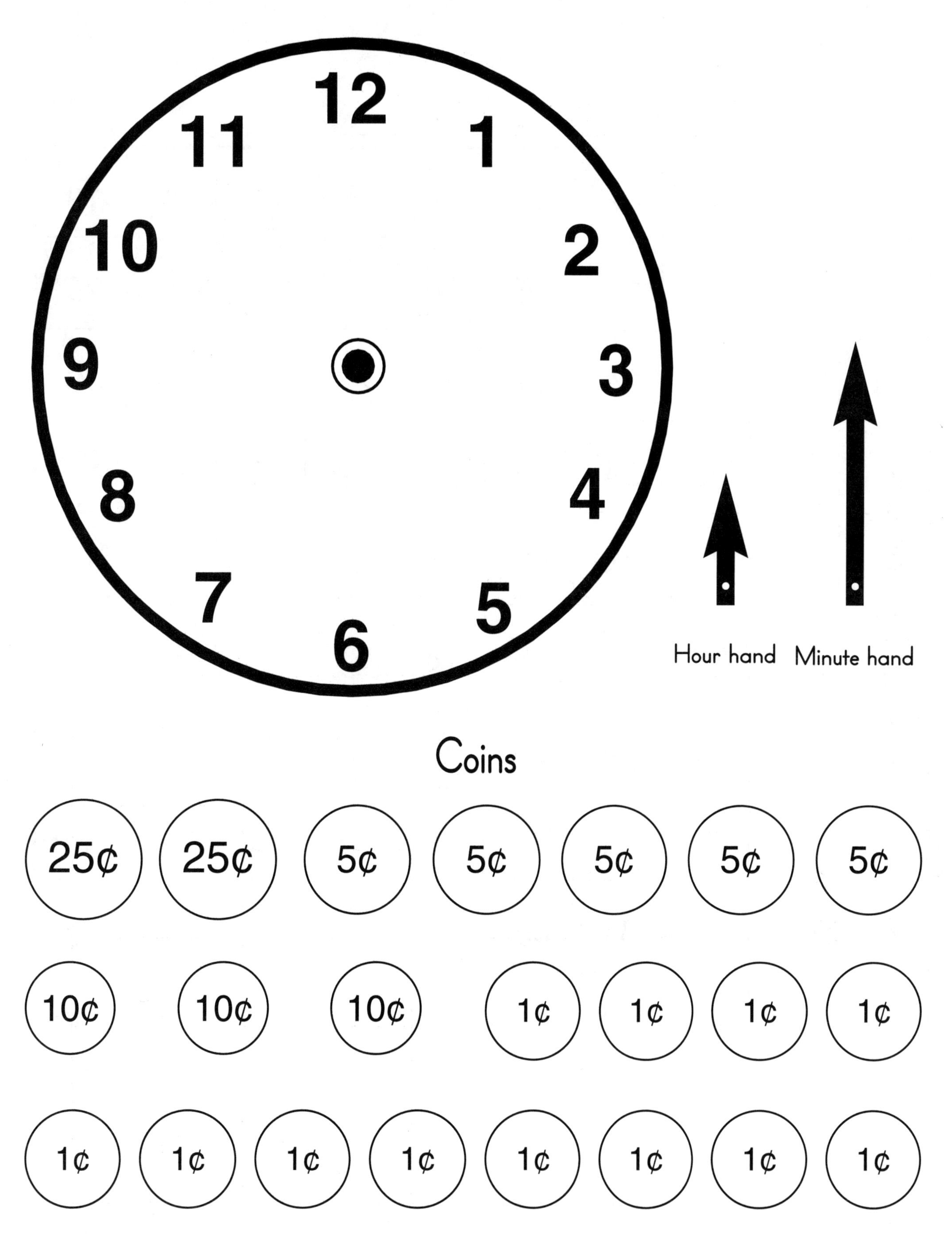

Hundreds Chart

1	2	3	4	5	6	7	8	9	10
11	12	13	14	15	16	17	18	19	20
21	22	23	24	25	26	27	28	29	30
31	32	33	34	35	36	37	38	39	40
41	42	43	44	45	46	47	48	49	50
51	52	53	54	55	56	57	58	59	60
61	62	63	64	65	66	67	68	69	70
71	72	73	74	75	76	77	78	79	80
81	82	83	84	85	86	87	88	89	90
91	92	93	94	95	96	97	98	99	100

Addition Chart

+	0	1	2	3	4	5	6	7	8	9	10
0	0	1	2	3	4	5	6	7	8	9	10
1	1	2	3	4	5	6	7	8	9	10	11
2	2	3	4	5	6	7	8	9	10	11	12
3	3	4	5	6	7	8	9	10	11	12	13
4	4	5	6	7	8	9	10	11	12	13	14
5	5	6	7	8	9	10	11	12	13	14	15
6	6	7	8	9	10	11	12	13	14	15	16
7	7	8	9	10	11	12	13	14	15	16	17
8	8	9	10	11	12	13	14	15	16	17	18
9	9	10	11	12	13	14	15	16	17	18	19
10	10	11	12	13	14	15	16	17	18	19	20

Multiplication Chart

X	0	1	2	3	4	5	6	7	8	9	10
0	0	0	0	0	0	0	0	0	0	0	0
1	0	1	2	3	4	5	6	7	8	9	10
2	0	2	4	6	8	10	12	14	16	18	20
3	0	3	6	9	12	15	18	21	24	27	30
4	0	4	8	12	16	20	24	28	32	36	40
5	0	5	10	15	20	25	30	35	40	45	50
6	0	6	12	18	24	30	36	42	48	54	60
7	0	7	14	21	28	35	42	49	56	63	70
8	0	8	16	24	32	40	48	56	64	72	80
9	0	9	18	27	36	45	54	63	72	81	90
10	0	10	20	30	40	50	60	70	80	90	100

